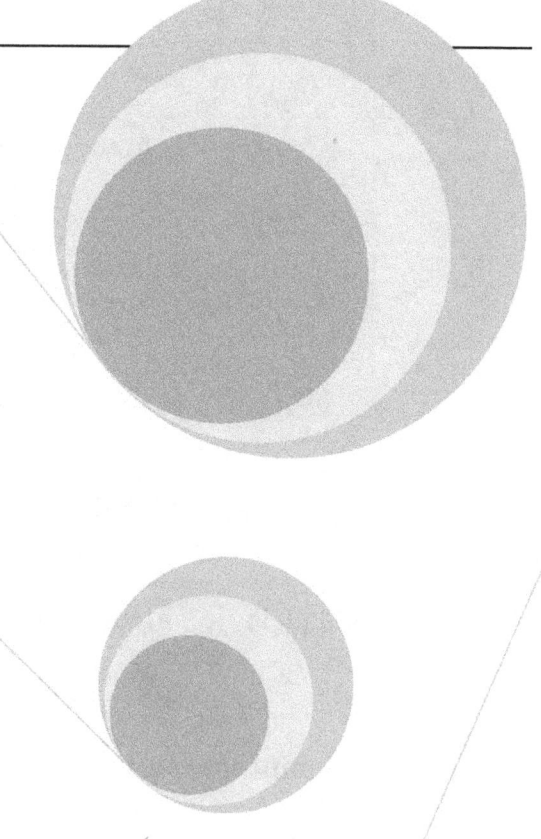

Project Management Skills for All Careers

By Project Management Open Resources and TAP-a-PM

Foreword by Daniel Dishno, Occupational Training Institute, De Anza College

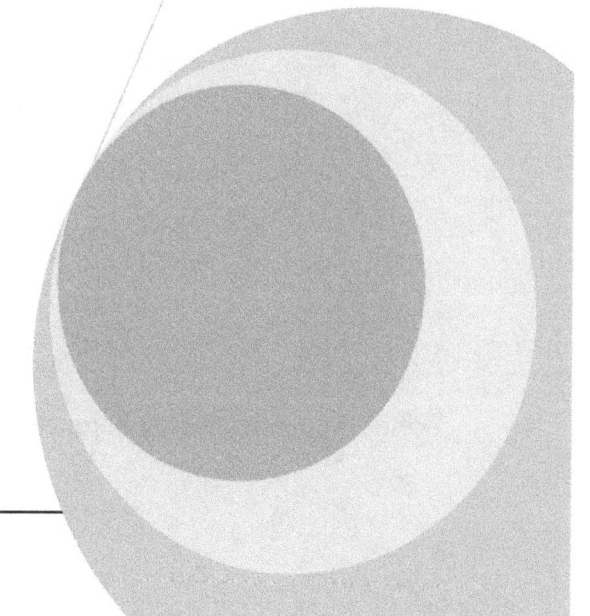

Bound Book
ISBN-13:978-1533322722
e book
ISBN-10: 0984813810
ISBN-13: 978-0-9848138-1-0

Based on Project Management for Scientists and Engineers by
Merrie Barron and Andrew Barron
http://cnx.org/content/col11120/1.4/

Significant contributions from the following sources

Maura Irene Jones, Career Descriptions in Chapter 1
Several photographs Copyright © 2011 by Maura Irene Jones
Creative Commons Attribution 3.0 CC BY
Attribution URL http://www.linkedin.com/in/maurajones

Randy Fisher, Chapter 12
(a subset of Organization Management and Development at http://wikieducator.org/OMD/Culture_PM)

Rekha Raman, Microsoft Word template and formatting

Ali Daimee, syllabus and more

Mike Milos, syllabus

Shuly Cooper, reviews

Bob Sawyer, Project Manager for the Saylor Foundation proposal

Victor Cesena, Project Manager for the bound textbook

Jim Huether, Program Manager

Jacky Hood, Managing Editor

See Appendix C for references used in the Barron & Barron book and in this book.

Project Management Open Resources (PMOR) is an organization dedicated to creating, publicizing, and distributing open-licensed project management information. PMOR's community includes over 60 members, many with extensive project management experience and certifications. See http://projectmanagementopenresources.ning.com/.

Tap-a-PM is a cooperative association of project and program managers founded in February of 2008 that acts as a source of accomplished program and project managers with full project life-cycle skills across a set of disciplines and industries. The association supports its members with a wealth of domain expertise and connections into the wider project and program management community. Tap-a-PM members include highly-qualified project and program managers with over 200 years of project/program management experience with backgrounds in software, computers, electronics, IT, on-line learning, instrumentation, telecom, bio-tech, and more. See http://www.tapapm.org/.

Professor Andrew Barron co-authored Project Management for Scientists and Engineers. He is the Charles W. Duncan, Jr.— Welch Chair of Chemistry, a Professor of Materials Science, and the Associate Dean for Industry Interactions and Technology Transfer at Rice University. See http://chemistry.rice.edu/FacultyDetail.aspx?RiceID=585.

Merrie Barron, a Project Management Professional and Certified Scrum Master, co-authored Project Management for Scientists and Engineers. She teaches project management for science and engineering at Rice University. See http://www.linkedin.com/in/merriephinney.

Victor Cesena is a Project Management Professional (PMP) and Certified Scrum Master (CSM). His extensive project management experience includes stints at Electronics for Imaging, Read-Rite, Sumitomo, Hitachi Metals, Tri-media, Kennedy, Eikon, and Ampex. Victor holds an Electrical Engineering degree from the University of the Pacific.

Shuly Cooper is president of PhytoScience, Inc. She has extensive experience in software quality engineering at companies including Verano, Spyglass, Space Systems Loral, Micro Focus, and Sybase. Shuly holds a Masters in Biochemistry from the Hebrew University and a PhD in Biophysics from the Weizmann Institute of Science.

Ali Daimee recently completed certification in program management at the University of California at Santa Cruz. He has taught project management at the University of California – Irvine, and Northwestern Pacific University. In his career Ali has worked for companies such as Broadcom, Sega, Sun, Oracle, Tandem, Compaq, Novell, Novellus, Cadence, Nortel, Control Data, Honeywell, and ICL. He has also co-founded startup companies managing development of web and network application products. Ali holds honors degrees in Mathematics and Electrical Engineering from London University.

Madhurika Dev, a project management consultant, has a track record of successfully managing software development projects through all phases of the project lifecycle. She has worked for FieldDay Solutions, Sourcecorp HealthServe, Cisco, HP and Sabre. Madhurika holds an MS in Mathematics from the University of Texas at Arlington.

Randy Fisher holds certifications in advanced technology management and instructional design. He is the Manager of Community Service Learning at the University of Ottawa's Centre for Global and Community Engagement. He has made significant contributions to the Community College Consortium for Open Education Resources, the Commonwealth of Learning, the OER Foundation and WikiEducator. Randy holds an MA in Organization Management and Development and a post-graduate degree in Journalism,

Jacky Hood is a program manager, service/support executive, management consultant, and educator. Prior to serving as Director of College Open Textbooks, her clients included Apple, HP, IBM, RightWave, and Slam Dunk Networks. She has published four books and numerous articles, and won writing awards from McGraw-Hill and Patton Consultants. Jacky holds a Masters of Systems Engineering from Carleton University, Ottawa, and a Bachelors degree in Electrical Engineering from the University of Nebraska.

Jim Huether, a Certified Scrum Master, has served as Program/Project Manager for companies such as Philips Semiconductors, Logitech, and Symantec, as well as Foothill College and College Open Textbooks, where he led the development of several on-line courses. Jim has also taught several courses in project management. Prior to this, he served in software development management positions for a number of companies before founding Nchant, an invention, licensing and product development company. Jim holds both a Bachelors degree and a Masters degree in Electrical Engineering from Rice University.

Mike Milos is a senior consultant at Deloitte & Touche, and teaches a system development lifecycle class at both undergraduate and graduate levels at the University of Phoenix In the past he has worked for Hewlett-Packard, Network Appliance, KLA-Tencor, and the US Navy. Mike holds a Masters degree in Computer Information Systems and a Bachelors degree in Information Technology from the University of Phoenix.

Maura Jones is a Project Management Professional, and holds certifications in audit (ISACA CISA) and security (ISC2 CISSP). Maura has provided project management expertise to a variety of global clients, and taught Business Data Communications and eCommerce at the University of San Francisco and Notre Dame de Namur University. Maura holds an MS in Telecommunications Management from Golden Gate University, a BS in Psychology from San Jose State University, and Certificates in Project Management from UC Berkeley and Stanford. Maura is active in professional organizations, including PMI and ITIL.

Rekha Raman, a Project Management Professional, is a marketing communications manager at LitePoint, responsible for a wide spectrum of documents, including datasheets, quick start guides, user manuals, field service instructions, and regulatory documents. With over 15 years of experience in technical writing, Rekha's interests range from effective communications to marine biology to wireless technology. On behalf of College Open Textbooks, she reviewed the Project Management for Scientists and Engineers textbook.

Lalit Sabnani is APICS certified and is working on his Project Management Professional certification. He has led large and complex development programs across a 25-year career in data storage and semiconductor technology. On behalf of College Open Textbooks, he reviewed the Project Management for Scientists and Engineers textbook. Lalit holds an MS in Industrial Engineering from Arizona State University and a BS in Mechanical Engineering from MS University in India.

Bob Sawyer, a consultant in product management and product marketing, has worked for a wide range of technology companies, both large and small, including IBM, Solid, Panta, HP, Compaq, and Tandem. He holds a Bachelors degree from Northwestern University and a Masters degree from the Kellogg School of Management.

Dalvinder Singh Matharu has worked as a project manager for 3 years and has been a team member for more than 15 years. He is a Project Management Professional (PMP) and a Certified Scrum Master (CSM).

Daria Hemmings holds an MA in Creative Writing from Emerson College and a Certificate in Systems Analysis from Northeastern University. She has taught Freshman Composition at Emerson College and Craven Community College..

Edition 3 Updates
1. Updated ISBN with new publisher (CreateSpace)
2. Modifed PDF security settings to prevent printing and modifications.
3. Created new cover for Createspace publishing.

Foreword

Daniel Dishno, Supervisor, Occupational Training Center,
De Anza College, Cupertino, CA, USA

Every organization has a purpose for existing. It has a set of ongoing organized functions and structures (aka work) that have been established to accomplish something that relates to the purpose of the organization. At a college, instructors teach classes, counselors provide academic advice, and administrators guide the day to day operations. This is not project management, it is ongoing work. *Project Management Skills for All Careers* defines project management as "the application of knowledge, skills, tools, and techniques applied to project activities in order to meet project requirements. Project management is a process that includes planning, putting the project plan into action, and measuring progress and performance. Projects are unique, temporary in nature and have a definite beginning and end. Projects are completed when the project goals are achieved. A successful project is one that meets or exceeds the expectations of the stakeholders."

In our department at the college, we have projects. Mainly these projects originate from grants and contracts. In developing a grant proposal or contract, team members gather around and look at what is required in the grant and try and figure out how best to submit a competitive proposal. Sometimes we are overwhelmed when we read what is expected. Sometimes we laugh, and we face our fears and proceed into the unknown. Our projects usually involve job training and job placement services, catering to a specified group of unemployed clients such as refugees, laid-off high tech workers, or welfare recipients. After reading *Project Management Skills for All Careers*, I realized the project plan is the same as a grant proposal or contract.

This is exciting stuff. We are bringing something new to the campus. A new group of students that just arrived from some war-torn country training for a job, a laid-off worker re-training for a new career, a welfare mom gaining skills to be self-sufficient, a new skills training program.

I read *Project Management Skills for All Careers* in less than one week, and I feel better equipped in planning, implementing, measuring, changing and completing projects. This book refreshed my passion for my work. I am excited to have this book in my arsenal of professional resources.

Project Management Skills for All Careers offers a framework for managing projects in any career area. The concepts can be applied no matter where you work. As a matter of fact, many of our dislocated workers are trained to become certified in Project Management. Project management skills are essential and invaluable for anyone who initiates or is assigned to a project. *Project Management Skills for All Careers* is a unique book, as it is current, well organized, a pleasure to read. It is available as an open source textbook, free to those who use and apply it in their work place.

When an opportunity presents itself, we look around for people with these skills: leadership, decisiveness, scoping, identifying tasks and deliverables, defining relationships among tasks, finding and assigning resources, scheduling, and budgeting. We also want soft skills including building relationships, communicating with all concerned parties, and motivating people to produce quality work quickly and efficiently.

Similarly when confronted with a problem such as a natural disaster, many of the same skills are required.

For more than half a century, project managers have learned and applied these skills in engineering, science, construction, and more. Today's rapidly-changing world calls for expanding the use of project management skills to many more industries and careers.

Managing repetitive work, process management, was the norm for centuries. Agriculture, manufacturing, retail, transportation, and other endeavors remained the same for years or decades. Those days are past. The world is moving much faster and all processes must change often. Changing a process is a project and it demands project management skills. No longer can a business manager, nurse, teacher, or any other worker assume that he or she can learn a routine and then repeat it for years.

The mission of this textbook is two-fold:

- To provide students with project management skills they can apply in any chosen profession.

- To provide instructors with an open-licensed textbook they can freely copy, move into a learning management system; and modify to suit their teaching style, student demographics, available teaching time, and more.

With attribution to the original authors Merrie Barron and Andrew Barron, the Project Management Open Resources community, the TAP-a-PM project/program management cooperative, and other sources, any instructor, indeed any person or organization, may freely use and even sell the materials in this textbook. Please include the information on the copyright page in your attribution.

Our project team invites all users of this textbook to learn, have fun, and be successful in their chosen careers.

The following syllabus is suggested for an introductory 15-week one-semester class in project management for business school students. For shorter terms such as 12-week quarters or multi-day workshops in industry, chapters 1-6 could be covered in a single session, and chapters 17 and 18 omitted and saved for an advanced class.

This textbook could also be used in many vocational programs; examples appear in Chapter 1. The particular skills needed in those occupations could be addressed, e.g., scheduling and budgeting.

Week/ Session #	Book Part #	Book Chapter #	Topic Covered	Assignment
1	I	1, 2, 3, 4, 5	Definition and characteristics of Project; Project Management and its history; Various applications Project Management and its benefits to business; Participants in Project Management its beneficiaries	Form small teams of 3-5 students; Brainstorm about a specific business the team wants to select and define a project for your team
2		6	Skill set and expertise necessary for a successful Project Manager; Examples and Challenges faced by a Project Manager; Focus on Interpersonal skills – the most important tool set	Practice interpersonal skills among your team members using role play and recognize leadership traits of your team
3		7	The Project Life Cycle and its phases – key activities, focus, and challenges of each phase	Define key deliverables per each phase for your teams project and define beginning and end of these phases for your project
4	II	8, 9, 10, 11, 12	Recognizing stakeholders, Project Political Environment, Organizational Culture and their importance in Project Initiation; Types of Project Management Certification and their benefits	Define your team's project environment, stakeholders, organizational culture, policies and initiate your project

Week/ Session #	Book Part #	Book Chapter #	Topic Covered	Assignment
5	III	13, 14	Inputs to Project Planning Phase, Factors considered during the Scope planning step of Project Planning	Develop the scope of your teams project
6		15	Schedule Planning step – tools and techniques – types of schedules and their characteristics; Activities, dependencies, relationships, graphical presentation, tracking, etc.	Develop a Work Breakdown Structure for your project, define activities and create a basic network diagram
7		16	Resource Planning step – Defining effort, durations and type of resources required for a project – types of estimates, tools used, adding information to the project schedule	Define resource needed for each activity, duration allowed and the effort required for your project – Update your project plan with this information
8		17	Budget Planning – Consideration of costs and tradeoffs of various execution options such as Company Internal cost of doing the project versus contracting or subcontracting all part of the project – developing a budget for the project	Develop a budget for your project considering a mix of subcontractors and internal resources
9		18	Risk and its definition; Risk identification process, Probability and impact consideration of Risks; Developing a Risk Register and identifying various Risk mitigation options.	Identify Risks on your project, their probability and impact, rank them and determine their triggers and mitigation options
10		19	Quality Planning considerations – Regulatory requirements, Industry standards, Internal Policies and guidelines, Quality monitoring and control, Quality Assurance and its benefits	Define a Quality plan for your project – consider Regulatory requirements, customer satisfaction, etc.

Week/ Session #	Book Part #	Book Chapter #	Topic Covered	Assignment
11		20	Communications Planning – Defining communications channels, types of communications, amount of communications, Defining Interfaces with Internal and external stakeholders, consideration of conflicts and their resolution, etc.	Create a communications plan for your project employing techniques learned in this chapter
12		21	Completing the overall Project Planning as the final deliverable from the Project Planning Phase	Review your overall project plan and optimize it as necessary
13	IV	22	Project Implementation Phase and its tracking and control – Need for replanning as and when needed; tools and techniques of Project control	Define change control plan for your project
14		23	Project Completion and how to recognize it – various actions involved in closing a project – importance of lessons learned and the celebration	Identify closing actions for your project and document Lessons Learned.
15		Final	Final review and Team presentations	

COMPLETE TABLE OF CONTENTS

Author and Contributor Bios

Foreword

Preface

A Word to Business School and other Instructors

PART I - INTRODUCTION

Chapter 1: Jump-Start Any Career with Project Management Skills

    1.1   Careers Using Project Management Skills

    1.2   Business Owners

    1.3   Construction Manager

    1.4   Creative Services

    1.5   Educator

    1.6   Engineers

    1.7   Healthcare Careers

    1.8   Paralegal

    1.9   Software developer/computer programmer

    1.10   Scientist Technicians

Chapter 2: History of Project Management

Chapter 3: What is a Project?

    3.1   A Formal Definition of a Project

Chapter 4: Project Characteristics

Chapter 5: What is Project Management?

Chapter 6: Project Management Areas of Expertise

    6.1   Application knowledge; standards & regulations

    6.2   Understanding the Project Environment

    6.3   Management Knowledge and Skills

6.4   Interpersonal Skills

Chapter 7: The Project Life Cycle

7.1   Initiation Phase

7.2   Planning Phase

7.3   Implementation Phase

7.4   Closing Phase

PART II – PROJECT STRATEGY

Chapter 8: Project Stakeholders

8.1   Top Management

8.2   The Project Team

8.3   Your Manager

8.4   Peers

8.5   Resource Managers

8.6   Internal Customer

8.7   External customer

8.8   Government

8.9   Contractors, subcontractors, and suppliers

Chapter 9: The Politics of Projects

9.1   Assess the environment

9.2   Identify goals

9.3   Define the problem

Chapter 10: Project Initiation

Chapter 11: Project Management Certifications

11.1   Project Management Institute Overview

11.2   Scrum Development Overview

Chapter 12: Culture and Project Management

12.1   What is Organizational Culture?

12.2   Project Manager's Checklist

12.3   Project Team Challenges

12.4   Dealing with Conflict

12.4   Bibliography for Chapter 12

PART III – PROJECT PLANNING

Chapter 13: Overview of Project Planning

Chapter 14: Scope Planning

14.1   Defining the Scope

14.2   Project Requirements

14.3   Functional Requirements

14.4   Non-Functional Requirements

14.5   Technical Requirements

14.6   User Requirements

14.7   Business Requirement

14.8   Regulatory requirements

14.9   An Example of Requirements

Chapter 15: Project Schedule Planning

15.1   Preparing the work breakdown structure

15.2   A case study

15.3   Activity Definition

15.4   Leads and Lags

15.4   Milestones

15.5   The Activity Sequencing Process

15.6   Creating the Network Diagram

Chapter 16: Resource Planning

16.1   Estimating the Resources

16.2   Estimating Activity Durations

16.3   Project Schedule

Chapter 17: Budget Planning

17.1   Make or Buy Analysis

17.2   Contract Types

Chapter 18: Risk Management Planning

Chapter 19: Quality Planning

19.1   Quality planning tool

Chapter 20: Communication Planning

Chapter 21: Bringing it all together

Part IV - IMPLEMENTATION and CLOSING

Chapter 22: Project Implementation Overview

22.1   Change control

Chapter 23: Project Completion

23.1   Lessons learned

23.2   Contract closure

23.3   Releasing project team

23.4   Celebrate!

Appendix A: Solutions to Exercises

Solution to Exercise 10.1

Solution to Exercise 15

Solutions to Exercises in Chapter 16

Appendix B: Glossary of Project Management Terms

Appendix C: Attributions and Bibliography

PART I - INTRODUCTION

New Moon Copyright ©TallPomlin CC BY
http://www.flickr.com/photos/paultomlin/364907230/sizes/m/in/photostream/

1.1   Careers Using Project Management Skills:

Skills learned by your exposure to studying project management can be used in most careers as well as in your daily life. Strong planning skills, good communication, ability to implement a project to deliver the product or service while also monitoring for risks and managing the resources, will provide an edge towards your success. Project Managers can be seen in many industry sectors including: Agriculture and Natural Resources; Arts, Media and Entertainment; Building Trades and Construction; Energy and Utilities; Engineering and Design; Fashion and Interiors; Finance and Business; Health and Human Services; Hospitality, Tourism and Recreation; Manufacturing and Product Development; Public and Private Education Services; Public Services; Retail and Wholesale Trade; Transportation; and Information Technology.

Below we explore various careers and some of the ways in which project management knowledge can be leveraged.

1.2   Business Owners

Business owners definitely need to have some project management skills. With all successful businesses, the product or service that is being delivered to the customer meets their needs in many ways. The product or service is of the quality desired, the costs are aligned with what the consumer expected, and the timeliness of that product or service meets the deadline for the buyer of that item.

Copyright © 2011 by Maura Irene Jones
Creative Commons Attribution 3.0 CC BY
Attribution URL http://www.linkedin.com/in/maurajones

The pillars of project management are delivering a product/service within schedule, cost, scope, and quality requirements. Business owners need planning, organizing, and scoping skills and the ability to analyze, communicate, budget, staff, equip, implement and deliver.

Understanding the finances, the operations, and the expenses of the business are among the skills that project managers learn and practice. Some businesses may focus more on accounting, providing financial advice, sales, training, public relations, and actuary or logistician roles. Business owners may own a travel agency or could provide hospitality. Business owners could be managing a store front or a location in their town's marketplace.

1.2.1   Example: Restaurant Owner/Manager

Restaurant Managers are responsible for the daily operations of a restaurant that prepares and serves meals and beverages to the customers. Strong planning skills, especially coordinating with the various departments (kitchen, dining room, banquet operations, food service managers, vendors providing the supplies) ensure that customers are satisfied with their dining experience. Managers' ability to recruit and retain employees, and monitor employee performance and training ensure quality with cost containment. Scheduling in many aspects, not only the staff but also the timing of the food service deliveries, is critical in meeting customer expectations.

Risk management is essential to ensure food safety and quality. Managers monitor orders in the kitchen to determine where delays may occur, and they work with the chef to prevent these delays. Legal compliance is essential in order for the restaurant to stay open, so Restaurant Managers direct the cleaning of the dining areas and the washing of tableware, kitchen utensils, and equipment. They ensure the safety standards and legality, especially in serving alcohol. Sensitivity and strong communication skills are needed when customers have complaints or employees feel pressured because more customers arrive than the forecast predicted.

Financial knowledge is needed for the soundness of running the restaurant, especially tracking special projects, events, and costs for the various menu selections. Catering events smoothly can be an outcome of using project plans and the philosophy of project management. The Restaurant Managers or the executive chef analyzes the recipes to determine food, labor, and overhead costs, determine the portion size and nutritional content of each serving, and assigns prices to various menu items, so that supplies can be ordered and received in time.

Planning is the key for successful implementation. Managers or executive chefs need to estimate food needs, place orders with distributors, and schedule the delivery of fresh food and supplies. They also plan for routine services (equipment maintenance, pest control, waste removal) and deliveries including linen services or the heavy cleaning of dining rooms or kitchen equipment, to occur during slow times or when the dining room is closed. A successful restaurant relies on many skills that the project management profession emphasizes.

Many businesses may explore outsourcing for certain services. Below is a sample status and project plan that reflects the various tasks needed for the project. A review of finances, the importance of communicating to stakeholders, and the importance of time, cost, schedule, scope, and quality are reflected. Many companies may use these steps in their business. These plans show the need for the entire team to review the various proposals to choose the best plan.

Sample status chart which is typical with the use of a red-yellow-green

Sample project plan exploring outsourcing of services

1.3   Construction Manager

Construction managers plan, direct, coordinate, and budget a wide variety of residential, commercial, and industrial construction projects including homes, stores, offices, roads, bridges, wastewater treatment plants, schools, and hospitals. Strong scheduling skills are essential for this role. Communication skills are often used in coordination of design and construction processes, teams executing the work and governance of special trades (carpentry, plumbing, electrical wiring) as well as government representatives for the permit processes.

The Construction Manager may be called a project manager or project engineer. The Construction Manager ensures that the project gets completed on time and within budget while meeting quality specifications and codes and maintaining a safe work environment. These managers create project plans in which they divide all required construction site activities into logical steps, estimating and budgeting the time required to meet established deadlines, usually utilizing sophisticated scheduling and cost-estimating software. Many use software packages such as Microsoft Project® or Procure® or online tools like BaseCamp®. Most construction projects rely on spreadsheets for project management. Procurement skills used in this field include acquiring the bills of material, lumber for the house being built, and more. Construction managers also cording labor, determining the needs and overseeing their performance, ensuring that all work is completed on schedule.

Values including sustainability, reuse, LEED-certified building, incorporating green energy, and various energy efficiencies are being incorporated into today's future projects. Ms. Jennifer Russell, spoke about "Project Management and Global Sustainability" at the 2011 Silicon Valley Project Management Institute (PMI) conference. She informed the attendees of the financial, environmental, and social areas in expanding the vision of project management with the slide shown here. These values are part of the PMI's Code of Ethics and professionalism in which the project manager includes in their decisions the best interest of society, the safety of the public, and enhancement of the environment.

1.4   Creative Services

Creative service careers include graphic artists, curators, video editors, gaming managers, multimedia artists, media producers, technical writers, interpreter, and translators. These positions use project management skills, especially in handling the delivery channel and meeting clients' requirements.

Let us look at one example, graphic artists, to understand and identify some of the project management skills that aid in this career.

1.4.1   Graphic Artists

Graphic artists plan, analyze, and create visual solutions to communications problems. They use many skills found in project management, especially communications. They work to achieve the most effective way to get messages across in print and electronic media. They emphasize their messages using color, type, illustration, photography, animation, and various print and layout techniques. Results can be seen in magazines, newspapers, journals, corporate reports, and other publications. Other deliverables from Graphic Artists using project management skills include promotional displays, packaging, and marketing brochures supporting products and services, logos, and signage. In addition to print media, graphic artists create materials for the web, TV, movies, and mobile device apps.

Initiation in project management can be seen in developing a new design: determining the needs of the client, the message the design should portray, and its appeal to customers or users. Graphic designers consider cognitive, cultural, physical, and social factors in planning and executing designs for the target audience, very similar to some of the dynamics a project manager considers in communicating with various project stakeholders. Designers may gather relevant information by meeting with clients, creative staff, or art directors; brainstorming with others within their firm or professional association, and by performing their own research to ensure that their results have high quality and to manage risks.

Graphic designers may supervise assistants who follow instructions to complete parts of the design process. Therefore scheduling, resource planning, and cost monitoring are pillars of project management seen in this industry. These artists use computer and communications equipment to meet their clients' needs and business requirements in a timely and cost-efficient manner.

1.5   Educators

_Educator' is a broad term that can describe a career in teaching, maybe being a lecturer, a professor, a tutor, or a home-schooler. Other educators include gurus, mullahs, pastors, rabbis, and priests. Instructors also provide vocational training or teach skills like learning how to drive a car or use a computer. Educators provide motivation to learn a new language or showcase new products and services. Educators use project management skills including planning and communication.

Let us look at a teacher since we all have had teachers and see if we can recognize the project management skills that are demonstrated in this profession.

1.5.1 Teachers

Some teachers foster the intellectual and social development of children during their formative years; other teachers provide knowledge, career skill sets and guidance to adults. Project management skills that teachers exhibit include acting as facilitators or coaches, communicating in the classroom and in individual instruction. Project managers plan and evaluate various aspects of a project; teachers also plan, evaluate, and assign lessons; implement these plans, and monitor student's progress similar to the way a project manager monitors and delivers goods or services. Teachers use their people skills to manage students, parents, administrators. The soft skills that project managers exercise can be seen in teachers encouraging collaboration in solving problems by having students work in groups to discuss and solve problems as a team.

Project Managers may work in a variety of fields with a broad assortment of people, similar to teachers who work with students from varied ethnic, racial, and religious backgrounds with awareness and understanding of different cultures.

Teachers in some schools may be involved in making decisions regarding the budget, personnel, textbooks, curriculum design, and teaching methods demonstrating skills that a project manager would possess such as finance, and decision making.

1.6 Engineers

Engineers apply the principles of science and mathematics to develop economical solutions to technical problems. As a project cycles from an idea in the project charter to the implementation and delivery of a product or service, engineers link scientific discoveries to commercial applications that meet societal and consumer needs.

Engineers use many project management skills, especially when engineers specify the functional requirements. Quality is observed in engineers as they evaluate the design's overall effectiveness, cost, reliability, and safety similar to the project manager reviewing the criteria for the customer's acceptance of delivery of the product or service.

Estimation skills in project management are used in Engineering. Engineers are asking many times to provide an estimate of time and cost required to complete projects.

1.7   Healthcare Careers

There are many jobs and careers in healthcare which use project management skills. The field of healthcare varies widely, such as athletic trainer, dental hygienist; massage therapist, occupational therapist, optometric, physician assistant and X-ray technicians. Again, these folks actively apply risk management in providing health care delivery of service to their clients, ensuring that they do not injury the person that they are caring for. Note: A section on nursing is covered within this area of the textbook.

Many of you may have experience taking a fall while you were growing up, and needed an x-ray to determine if you had a fracture or merely a sprain. Hence let us look at this career as an example of a healthcare professional using project management skills.

1.7.1   Radiologic Technologists and Technicians

Radiologic technologists and technicians perform diagnostic imaging examinations like x-rays, computed tomography, magnetic resonance imaging, and mammography. They could also be called radiographers, because they produce x-ray films (radiographs) of parts of the human body for use in diagnosing medical problems.

Project management skills, especially people skills and strong communication, are demonstrated when they prepare patients for radiologic examinations by explaining the procedure and what position the patient needs to be at, so that the parts of the body can be appropriately radiographed. Risk management is demonstrated when these professionals work to prevent unnecessary exposure to radiation, these workers surround the exposed area with radiation protection devices, such as lead shields, or limit the size of the x-ray beam. Quality is needed to provide the expected results, with the health technician monitoring the radiograph and setting controls on the x-ray machine to produce radiographs of the appropriate density, detail, and contrast.

Safety and regulations concerning the use of radiation to protect themselves, their patients, and their coworkers from unnecessary exposure is tracked in an efficient manner and reported as a control to ensure compliance. Project management skills can also be use for they may prepare work schedules, evaluate purchases of equipment, or manage a radiology department.

Some radiologic technologists specialize in computed tomography (CT), as CT technologists as they too use project management skills. Since CT scans produce a substantial amount of cross-sectional x rays of an area of the body, the CT uses ionizing radiation; therefore, it requires the same precautionary measures that are used with x rays, hence the need for risk management and monitoring for exposure.

Teamwork, not only with the patient which the Radiologic technologist is supporting, the doctor whom ordered the request, but also other healthcare providers rely on strong communication, quality, work done in a timely manner and using the hospital resources wisely boil down to ensuring that the project management triangle of cost, schedule, scope with quality delivered remain the essentials which provide a cornerstone to project management and the skills needed to obtain the objective.

1.7.2   Nurse

Nurses treat and educate patients and their family and public about various medical conditions and provide advice and emotional support. Nurses establish a care plan for their patients, activities like scheduling administering of medications as well as discontinuation of meds, i.e. intravenous (IV) lines for fluid, medication, blood, and blood products; applying therapies and treatments. Communication with the patient, their family, physicians and other healthcare clinicians may be done in person, or could use technology. Telehealth allows personnel to provide care and advice through electronic communications media including videoconferencing, the Internet, or by telephone.

Risk management is very important for a nurse, with some cases having a life or death consequence! Monitoring of pain management and vital signs and providing status to physicians help in responding to the health care needs of the patient.

The nursing field varies. Some nurses work in Infection control. They identify, track, and control infectious outbreaks in healthcare facilities and create programs for outbreak prevention and response to biological terrorism. Others are Educators, nurses who plan, develop, execute and evaluate educational programs and curricula for the professional development of students and professional nurses. Nurses may use project management skills while conducting healthcare consultations, advising on public policy, researching in the field or providing sales support of a product or service.

1.8   Paralegal

Attorneys assume the ultimate responsibility for legal work but they often obtain assistance. Paralegals assume this role in law firms and perform many tasks to aid in the legal profession. However, they are explicitly prohibited from carrying out duties considered to be the practice of law (i.e. giving legal advice, setting legal fees, court case presentations).

Project management skills from such as planning are used in helping lawyers prepare for closings, hearings, trials, and corporate meetings. Communication skills are used when paralegals prepare written reports that help attorneys determine how cases should be handled, or the preparation of various drafts, such as pleading and motions to be filed, obtain affidavits, etc.

Monitoring tasks aid Paralegals who may track files of all important case documents, working on risk containment on filing dates and responses to the court. Procurement considerations , skills that a project manager holds, can also be seen from a paralegal perspective via negotiation terms of hiring expert witnesses as well as other services such as acquiring services from process servers.

Financial skills may be use as well, such as assisting in preparing tax returns, establishing trust funds, and planning estates or maintain financial office records at the law firm.

Government, litigation, personal injury, corporate law, criminal law, employee benefits, intellectual property, labor law, bankruptcy, immigration, family law, and real estate are many different law practices a Paralegal professional may experience which can use project management skills in these various work environments.

1.9   Software developer/computer programmer:

Computer software developers and computer programmers design and develop software. They apply the principles of computer science and mathematics to create, test, and evaluate software applications and systems that make computers come alive. Software is developed in many kinds of projects: computer games, business applications, operating systems, network control systems, and more. Project management skills help develop the requirements for the software, identify and track the product development tasks, team communications, test cases, and management of the quality, schedule and resources (staff, equipment, labs, and more).

1.10   Scientist Technicians

Science Technicians use principles and theories of science and mathematics to assist in research and development and to help invent and improve products and processes with their jobs more practically oriented than scientists. Planning skills project managers use can be seen as Science Technicians set up, operate, and maintain laboratory instruments, monitor experiments, observe, calculate and record results. Quality is a factor here as it is in Project Management, essential in work to ensure the processes performed correctly, with proper proportions of ingredients, for purity, or for strength and durability.

There are different fields in which these scientist technicians can apply project management skills. Agricultural and food science technicians work with the testing on food and other agricultural products, involved in food, fiber, and animal research, production, and processing. Control and risk management are important here in executing the tests and experiments to improve the yield and quality of crops, or plants and animals resistance to disease, insects, or other hazards. Quality factors are emphasis when food science technicians may conduct tests on food additives and preservatives to ensure compliance with Food and Drug Administration regulations regarding color, texture, and nutrients.

Soil chemistry—Toxins in rice plants

Biological technicians work with biologists studying living organisms. Many assist scientists who conduct medical research or who work in pharmaceutical companies help develop and manufacture medicines. Skills in schedule, especially in incubation periods for the study of the impact on cells, could impact projects, such as exploring and isolating variables for research in living organisms and infectious agents. Biotechnology technicians apply knowledge and execution skills, techniques, gained from basic research, including gene splicing and recombinant DNA, and apply them to product development. Project Managers skills can be seen in the collaboration and communication between the team to record and understand the results and progress towards a cure or product.

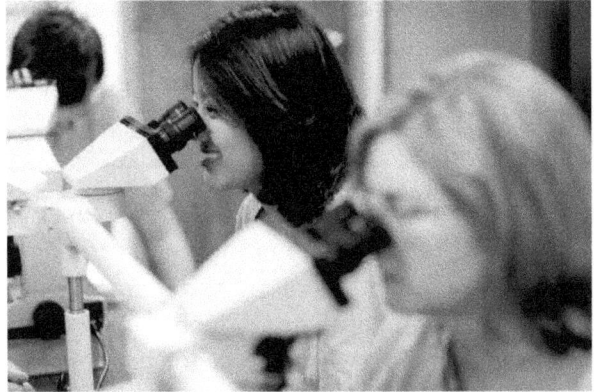

Photo provided by LAVA Pathology Specialists CC BY

Other kinds of technicians could be Chemical technicians who may work in laboratories and factories, using skills of monitoring and control in the way they collect and analyze samples. Again, quality assurance is of concern for most process technicians' work in manufacturing, testing packaging for design, integrity of materials, and environmental acceptability.

Technicians carry with them skills set from project management to assist in their initiation, planning, executing of their task, while managing risks with some measure of reporting to determine if their objectives are meet with the constraints of cost, schedule, resource, meeting quality standards set.

Could the Great Wall of China, the pyramids, or Stonehenge (Figure 2.1) have been built without project management? It is possible to say that the concept of project management has been around since the beginning of history. It has enabled leaders to plan bold and massive projects and manage funding, materials and labor within a designated time frame.

Figure 2.1: Stonehenge was erected between 3,000 BC and 1,600 BC by no less than three different cultures and its orientation on the rising and setting sun has always been one of its remarkable features

photo from Barron & Barron Project Management for Scientists and Engineers, http://cnx.org/content/col11120/1.4/

In late 19th century, in the United States, large-scale government projects were the impetus for making important decisions that became the basis for project management methodology such as the transcontinental railroad, which began construction in the 1860s. Suddenly, business leaders found themselves faced with the daunting task of organizing the manual labor of thousands of workers and the processing and assembly of unprecedented quantities of raw material.

Near the turn of the century, Frederick Taylor (Figure 2.2) began his detailed studies of work. He applied scientific reasoning to work by showing that labor can be analyzed and improved by focusing on its elementary parts that introduced the concept of working more efficiently, rather than working harder and longer.

Figure 2.2: Frederick Taylor (1856-1915).
photo from Barron & Barron Project Management for Scientists and Engineers, http://cnx.org/content/col11120/1.4/

Taylor's associate, Henry Gantt (Figure 2.4), studied in great detail the order of operations in work and is most famous for developing the Gantt Chart in the 1910s.

Figure 2.4 Henry Gantt (1861 -1919)
Photo from Barron & Barron Project Management for Scientists and Engineers, http://cnx.org/content/col11120/1.4/

A Gantt chart is a popular type of bar chart that illustrates a project schedule and has become a common technique for representing the phases and activities of a project work breakdown structure, so they can be understood by a wide audience (Figure 2.5). Although now considered a common charting technique, Gantt charts were considered quite revolutionary at the time they were introduced. Gantt charts were employed on major infrastructure projects including the Hoover Dam and the Interstate highway system and are still accepted today as important tools in project management.

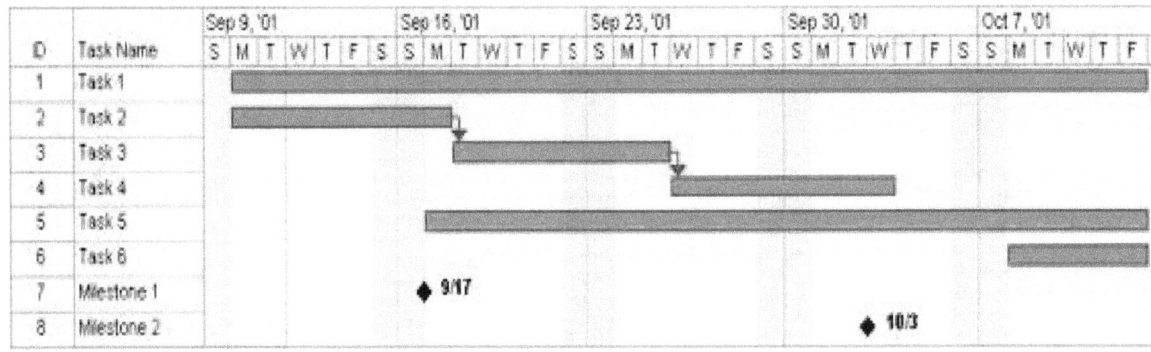

Figure 2.5: An example of a Gantt chart showing the relationship between a series of tasks.
Illustration from Barron & Barron Project Management for Scientists and Engineers, http://cnx.org/content/col11120/1.4/

By the mid Twentieth century, projects were managed on an ad hoc basis using mostly Gantt Charts, and informal techniques and tools. During that time, the Manhattan project was initiated and its complexity was only possible because of project management methods. The Manhattan project was the codename given to the Allied effort to develop the first nuclear weapons during World War II. It involved over thirty different project sites in the US and Canada, and thousands of personnel from US, Canada and UK. Born out of a small research program that began in 1939, the Manhattan Project would eventually employ 130,000 people and cost a total of nearly 2 billion USD and result in the creation of multiple production and research sites operated in secret. The project succeeded in developing and detonating three nuclear weapons in 1945.

The 1950s marked the beginning of the modern Project Management era. Two mathematical project-scheduling models were developed:

The Program Evaluation and Review Technique (PERT) was developed by Booz-Allen & Hamilton as part of the United States Navy's Polaris missile submarine program. PERT is basically a method for analyzing the tasks involved for completing a given project, especially the time needed to complete each task, the dependencies among tasks, and the minimum time needed to complete the total project (Figure 2.6).

The Critical Path Method (CPM) developed in a joint venture by both DuPont Corporation and Remington Rand Corporation for managing plant maintenance projects. The critical path determines the float, or schedule flexibility, for each activity by calculating the earliest start date, earliest finish date, latest start date, and latest finish date for each activity. The critical path is generally the longest full path on the project. Any activity with a float time that equals zero is considered a critical path task. CPM can help you figure out how long your complex project will take to complete and which activities are critical; meaning they have to be done on time or else the whole project will take longer. These mathematical techniques quickly spread into many private enterprises.

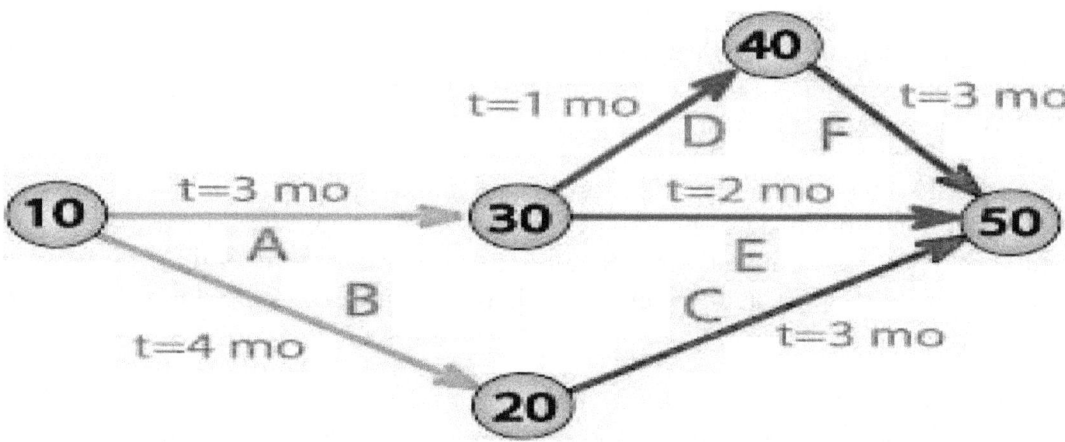

Figure 2.6: An example of a PERT network chart for a seven-month project with five milestones.
Illustration from Barron & Barron Project Management for Scientists and Engineers, http://cnx.org/content/col11120/1.4/

Project management in its present form began to take root a few decades ago. In the early 1960s, industrial and business organizations began to understand the benefits of organizing work around projects. They understood the critical need to communicate and integrate work across multiple departments and professions.

The starting point in discussing how projects should be properly managed is to first understand what a project is and just as importantly what it is not.

People have been undertaking projects since the earliest days of organized human activity. The hunting parties of our prehistoric ancestors were projects, for example; they were temporary undertakings directed at the goal of obtaining meat for the community. Large complex projects have also been with us for a long time. The pyramids and the Great Wall of China were in their day of roughly the same dimensions as the Apollo Project to send men to the moon. We use the term project frequently in our daily conversations. A husband, for example may tell his wife, "My main project for this weekend is to straighten out the garage." Going hunting, building pyramids, and fixing faucets all share certain features that make them projects.

A project has distinctive attributes, which distinguish it from ongoing work or business operations. Projects are temporary in nature. They are not an everyday business process and have definitive start dates and end dates. This characteristic is important because a large part of the project effort is dedicated to ensuring that the project is completed at the appointed time. To do this, schedules are created showing when tasks should begin and end. Projects can last minutes, hours, days, weeks, months or years.

Projects exist to bring about a product or service that hasn't existed before. In this sense, a project is unique. Unique means that this is new, this has never been done before. Maybe it's been done in a very similar fashion before but never exactly in this way. For example, Ford Motor Company is in the business of designing and assembling cars. Each model that Ford designs and produces can be considered a project. The models differ from each other in their features and are marketed to people with various needs. An SUV serves a different purpose and clientele than a luxury model. The design and marketing of these two models are unique projects. However the actual assembly of the cars is considered an operation, i.e., a repetitive process that is followed for most makes and models.

In contrast with projects, operations are ongoing and repetitive. They involve work that is continuous without an ending date and you often repeat the same processes and produce the same results. The purpose of operations is to keep the organization functioning while the purpose of a project is to meet its goals and to conclude. Therefore, operations are ongoing while projects are unique and temporary.

The project is completed when its goals and objectives are accomplished. It is these goals that drive the project and all the planning and implementation efforts undertaken to achieve them. Sometimes projects end when it is determined that the goals and objectives cannot be accomplished or when the product or service of the project is no longer needed and the project is cancelled.

3.1   A Formal Definition of a Project

There are many written definitions of a project. All of them contain the key elements described above. For those looking for a formal definition of a project, PMI defines a project as a temporary endeavor undertaken to create a unique product, service, or result. The temporary nature of projects indicates a definite beginning and end. The end is reached when the project's objectives have been achieved or when the project is terminated because its objectives will not or cannot be met, or when the need for the project no longer exists.

Chapter 4: Project Characteristics

When considering whether or not you have a project on your hands, there are some things to keep in mind. First, is it a project or ongoing operation? Next, if it is a project; who are the stakeholders? And third, what characteristics distinguish this endeavor as a project?

A project has several characteristics:

- Projects are unique.

- Projects are temporary in nature and have a definite beginning and ending date.

- Projects are completed when the project goals are achieved or it's determined the project is no longer viable.

A successful project is one that meets or exceeds the expectations of the stakeholders.

Consider the following scenario: The VP of Marketing approaches you with a fabulous idea. (Obviously it must be "fabulous" because he thought of it.) He wants to set up kiosks in local grocery stores as mini offices. These offices will offer customers the ability to sign up for car and home insurance services as well as make their bill payments. He believes that the exposure in grocery stores will increase awareness of the company's offerings. He told you that senior management has already cleared the project and he'll dedicate as many resources to this as he can. He wants the new kiosks in place in 12 selected stores in a major city by the end of the year. Finally, he has assigned you to head up this project.

Your first question should be "Is it a project?" This may seem elementary, but confusing projects with ongoing operations happens often. Projects are temporary in nature, have definite start and end dates, result in the creation of a unique product or service, and are completed when their goals and objectives have been met and signed off by the stakeholders.

Using these criteria, let's examine the assignment from the VP of marketing to determine if it is a project:

- Is it unique? Yes, because the kiosks don't exist in the local grocery stores. This is a new way of offering the company's services to its customer base. While the service the company is offering isn't new, the way it is presenting its services is.

- Does the product have a limited timeframe? Yes, the start date of this project is today, and the end date is the end of next year. It is a temporary endeavor.

- Is there a way to determine when the project is completed? Yes, the kiosks will be installed and the services will be offered from them. Once all the kiosks are intact and operating, the project will come to a close.

- Is there a way to determine stakeholder satisfaction? Yes, the expectations of the stakeholders will be documented in the form of requirements during the planning processes. These requirements will be compared to the finished product to determine if it meets the expectations of the stakeholder.

If the answer is yes to all these questions, then "Houston, we have a project".

You've determined that you have a project. What now? The notes you scribbled down on the back of the napkin at lunch are a start, but not exactly good project management practice. Too often, organizations follow Nike's advice when it comes to managing projects when they "just do it." An assignment is made and the project team members jump directly into the development of the product or service requested. In the end the delivered product doesn't meet the expectations of the customer. Unfortunately, many projects follow this poorly constructed path and that is a primary contributor to why a large percentage of projects don't meet their original objectives defined by performance, schedule, and budget.

In the United States, more than $250 billion dollars is spent each year on IT application development in approximately 175,000 projects. The Standish Group (a Boston-based leader in project and value performance research) released the summary version of their 2009 CHAOS Report that tracks project failure rates across a broad range of companies and industries (Figure 5.1).

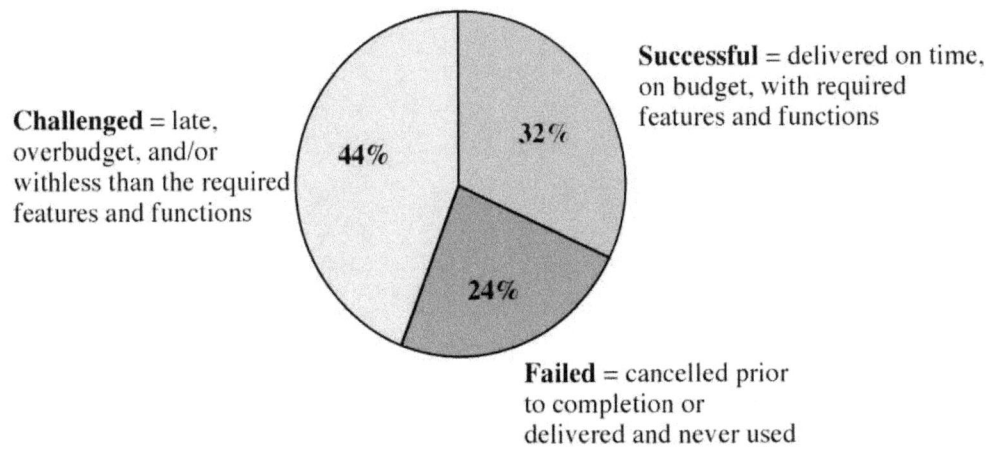

Figure 5.1: Summary of 2009 Standish Group CHAOS report.
Illustration from Barron & Barron Project Management for Scientists and Engineers, http://cnx.org/content/col11120/1.4/

Jim Johnson, chairman of the Standish Group has stated that "this year's results show a marked decrease in project success rates, with 32% of all projects succeeding which are delivered on time, on budget, with required features and functions, 44% were challenged which are late, over budget, and/or with less than the required features and functions and 24% failed which are cancelled prior to completion or delivered and never used."

When are companies going to stop wasting billions of dollars on failed projects? The vast majority of this waste is completely avoidable; simply get the right business needs (requirements) understood early in the process and ensure that project management techniques are applied and followed and the project activities are monitored.

Applying good project management discipline is the way to help reduce the risks. Having good project management skills does not completely eliminate problems, risks, or surprises. The value of good project management is that you have standard processes in place to deal with all contingencies.

Project Management is the application of knowledge, skills, tools, and techniques applied to project activities in order to meet the project requirements. Project management is a process that includes planning, putting the project plan into action, and measuring progress and performance.

Managing a project includes identifying your project's requirements; writing down what everyone needs from the project. What are the objectives for your project? When everyone understands the goal, it's much easier to keep them all on the right path. Make sure you set goals that everyone agrees on to avoid team conflicts later on. Understanding and addressing the needs of everyone affected by the project means the end result of your project is far more likely to satisfy your stakeholders, and last but not least, as project manager you will also be balancing the many competing project constraints.

On any project, you will have a number of competing project constraints that are competing for your attention. They are cost, scope, quality, risk, resources and time.

- Cost is budget approved for the project including all necessary expenses needed to deliver the project. Within organizations, project managers have to balance between not running out of money and not under spending because many projects receive funds or grants that have contract clauses with an "use it or lose it" approach to project funds. Poorly executed budget plans can result in a last minute rush to spend the allocated funds. For virtually all projects, cost is ultimately a limiting constraint; few projects can go over budget without eventually requiring a corrective action.

- Scope is what the project is trying to achieve. It entails all the work involved in delivering the project outcomes and the processes used to produce them. It is the reason and the purpose of the project.

- Quality is the standards and criteria to which the project's products must be delivered for them to perform effectively. First, the product must perform to provide the functionality expected, and to solve the problem, and deliver the benefit and value expected of it. It must also meet other performance requirements, or service levels, such as availability, reliability and maintainability, and have acceptable finish and polish. Quality on a project is controlled through quality assurance (QA) that is the process of evaluating overall project's performance on a regular basis to provide confidence that the project will satisfy the relevant quality standards.

- Risk is defined by potential external events that will have a negative impact on your project if they occur. Risk refers to the combination of the probability the event will occur and the impact on the project if the event occurs. If the combination of the probability of the occurrence and the impact to the project is too high, you should identify the potential event as a risk and put a proactive plan in place to manage the risk.

- Resources are required to carry out the project tasks. They can be people, equipment, facilities, funding, or anything else capable of definition (usually other than labor) required for the completion of a project activity.

- Time is defined as the time to complete the project. Time is often the most frequent project oversight in developing projects. This is reflected in missed deadlines and incomplete deliverables. Proper control of the schedule requires the careful identification of tasks to be performed, an accurate estimation of their durations, the sequence in which they are going to be done, and how people and other resources are allocated. Any schedule should take into account vacations and holidays.

You may have heard of the term "Triple Constraint" which traditionally only consisted of Time, Cost & Scope. These are the primary competing project constraints that you have to be aware of most. The triple constraint is illustrated in the form of a triangle to visualize the project work and to see the relationship between the scope/quality, schedule/time, and cost/resource (Figure 5.2).

Figure 5.2: A schematic of the triple constraint triangle.
Illustration from Barron & Barron Project Management for Scientists and Engineers, http://cnx.org/content/col11120/1.4/

In this triangle, each side represents one of the constraints (or related constraints) wherein any changes to any one side cause a change in the other sides. The best projects have a perfectly balanced triangle. Maintaining this balance is difficult because projects are prone to change. For example, if scope increases, cost and time may increase disproportionately. Alternatively, if the amount of money you have for your project decreases, you may be able to do as much, but your time may increase.

Your project may have additional constraints that you are facing, and as the project manager you have to balance the needs of these constraints against the needs of the stakeholders and against your project goals. For instance, if your sponsor wants to add functionality to the original scope you will very likely need more money to finish the project or if they cut the budget you have to reduce the quality of your scope and if you don't get the appropriate resources to work on your project tasks you will have to extend your schedule because the resources you have take much longer to finish the work.

You get the idea; they are all dependent on each other. Think of all of these constraints as the classic carnival game of Whac-a-mole (Figure 5.3). Each time you try to push one mole back in the hole, another one pops out. The best advice is to rely on your project team to keep these moles in place.

Figure 5.3: Go to www.dorneypark.com/public/online fun/mole.cfm to play Whac-a-mole.
Photo from Barron & Barron Project Management for Scientists and Engineers, http://cnx.org/content/col11120/1.4/

Here is an example of a project that cut quality because the project costs were fixed. The P-36 oil platform (Figure 5.4) was the largest footing production platform in the world capable of processing 180,000 barrels of oil per day and 5.2 million cubic meters of gas per day. Located in the Roncador Field, Campos Basin, Brazil the P-36 was operated by Petrobras.

Figure 5.4.: The Petrobras P-36 oil platform.
Photo from Barron & Barron Project Management for Scientists and Engineers, http://cnx.org/content/col11120/1.4/

In March 2001, the P-36 was producing around 84,000 barrels of oil and 1.3 million cubic meters of gas per day when it became destabilized by two explosions and subsequently sank in 3900 feet of water with 1650 short tons of crude oil remaining on board, killing 11 people. The sinking is attributed to a complete failure in quality assurance, and pressure for increased production led to corners being cut on safety procedures. It is listed as one of the most expensive accidents with a price tag of $515,000,000.

The following quote is from a Petrobras executive, citing the benefits of cutting quality assurance and inspection costs on the project, while the accompanying pictures are the result of this proud achievement in project management by Petrobras. The quotation is provided **one sentence at a time and compared with pictures of the actual outcome**

Figure 5.5: "Petrobras has established new global benchmarks for the generation of exceptional shareholder wealth through an aggressive and innovative program of cost cutting on its P36 production facility."
Photo from Barron & Barron Project Management for Scientists and Engineers, http://cnx.org/content/col11120/1.4/

Figure 5.6: "Conventional constraints have been successfully challenged and replaced with new paradigms appropriate to the globalized corporate market place."
Photo from Barron & Barron Project Management for Scientists and Engineers, http://cnx.org/content/col11120/1.4/

Figure 5.7: "Through an integrated network of facilitated workshops, the project successfully rejected the established constricting and negative influences of prescriptive engineering, onerous quality requirements, and outdated concepts of inspection and client control."
Photo from Barron & Barron Project Management for Scientists and Engineers, http://cnx.org/content/col11120/1.4/

Figure 5.8: "Elimination of these unnecessary straitjackets has empowered the project's suppliers and contractors to propose highly economical solutions, with the win-win bonus of enhanced profitability margins for themselves."
Photo from Barron & Barron Project Management for Scientists and Engineers, http://cnx.org/content/col11120/1.4/

Figure 5.9: "The P36 platform shows the shape of things to come in the unregulated global market economy of the 21st century."
Photo from Barron & Barron Project Management for Scientists and Engineers, http://cnx.org/content/col11120/1.4/

The dynamic trade-offs between the project constraint values have been humorously and accurately described in Figure 5.10.

"We can do GOOD, QUICK and CHEAP work.

You can have any two but not all three.

1. GOOD QUICK work won't be CHEAP.
2. GOOD CHEAP work won't be QUICK.
3. QUICK CHEAP work won't be GOOD."

Figure 5.10: A sign seen at an automotive repair shop.

Illustration from Barron & Barron Project Management for Scientists and Engineers, http://cnx.org/content/col11120/1.4/

In order for you as the project manager to manage the competing project constraints and to manage the project as a whole, there are some areas of expertise that you should bring onto the project team (Figure 6.1). They are the application area of knowledge; standards and regulations in your industry, understanding the project environment, and you must have general management knowledge and interpersonal skills. It should be noted that the industry expertise is not in a certain field but the expertise in order to run the project. So while knowledge of the type of industry is important you will have a project team supporting you in this endeavor. For example, if you are managing a project that is building an oil platform, you would not be expected to have a detailed understanding of the engineering since your team will have mechanical and civil engineers who will provide the appropriate expertise, however, it would definitely help if you understand this type of work.

Let's take a look at each of these areas in more detail.

6.1   Application knowledge; standards & regulations

By standards, we mean guidelines or preferred approaches that are not necessarily mandatory. In contrast, when referring to regulations we mean mandatory rules that must be followed such as Government imposed requirements through laws. It should go without saying that as a professional, you're required to follow all applicable laws and rules that apply to your industry, organization, or project. Every industry has standards and regulations. Knowing which ones affect your project before you begin work will not only help the project to unfold smoothly, but will also allow for effective risk analysis.

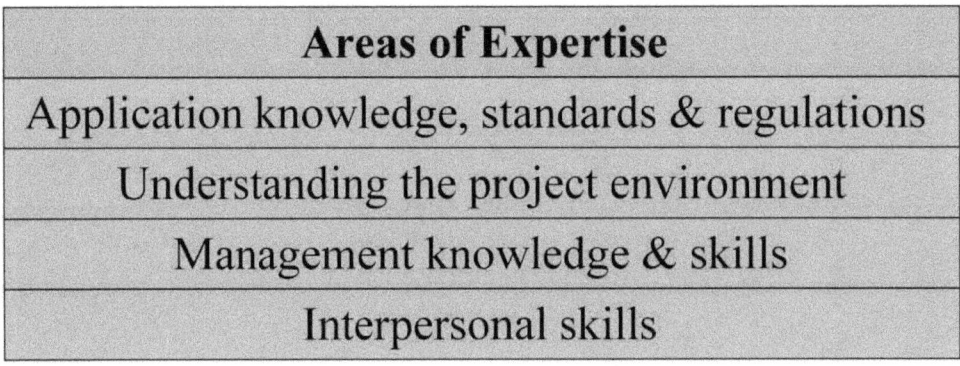

Areas of Expertise
Application knowledge, standards & regulations
Understanding the project environment
Management knowledge & skills
Interpersonal skills

Figure 6.1: Areas of expertise that a project manager should bring to the project team.
Table from Barron & Barron Project Management for Scientists and Engineers, http://cnx.org/content/col11120/1.4/

Some projects require specific skills in certain application areas. Application areas are made up of categories of projects that have common elements. They can be defined by: industry group (pharmaceutical, financial, etc), by department (accounting, marketing, legal, etc), by technical (software development, engineering, etc), or management (procurement, research, & development, etc) specialties. These application areas are usually concerned with disciplines, regulations and the specific needs of the project, the customer, or the industry. For example, most government agencies have specific procurement rules that apply to their projects that wouldn't be applicable in the construction industry. The pharmaceutical industry is interested in regulations set forth by the Food and Drug Administration, whereas the automotive industry has little or no concern for either of these types of regulations. You need to stay up-to-date regarding your industry so that you can apply your knowledge effectively. Today's fast paced advances can leave you behind fairly quickly if you don't stay abreast on current trends.

Having some level of experience in the application area you're working in will give you an advantage when it comes to project management. While you can call in experts who have the application area knowledge, it doesn't hurt for you to understand the specific aspects of the application areas of your project.

6.2   Understanding the Project Environment

There are many factors that need to be understood within your project environment (Figure 6.2). At one level you need to understand your project environment by thinking in terms of the cultural and the social environment. In this region we think of people, demographics and education. The international and political environment is where you need to understand about different countries cultural influences. Then we move on to the physical environment; here we think about time zones, think about different countries and how differently your project will be executed whether it is just in your country or whether you have an international project team that is distributed throughout the world in five different countries.

Project Environment	
Cultural	Social
International	Political
Physical	

Figure 6.2: The important factors to consider within the project environment.
Table from Barron & Barron Project Management for Scientists and Engineers, http://cnx.org/content/col11120/1.4/

Of all the factors the physical ones are the easiest to understand, and it is the cultural and international factors that are often misunderstood or ignored. How we deal with clients, customers, or project members from other countries can be critical to the success of the project. For example, the culture of the United States values accomplishments and individualism.

Americans tend to be informal and call each other by first names, even if having just met. Europeans tend to be more formal, using surnames instead of first names in a business setting, even if they know each other well. In addition, their communication style is more formal than in the US, and while they tend to value individualism, they also value history, hierarchy, and loyalty. The Japanese, on the other hand, tend to communicate indirectly and consider themselves part of a group, not as individuals. The Japanese value hard work and success, as most of us do.

How a product is received can be very dependent on the international cultural differences. For example, in the nineties, when many large American and European telecommunications companies were cultivating new markets in Asia, their customer's cultural differences often produced unexpected situations. Western companies planned their telephone systems to work the same way in Asia as they did in Europe and America. But the protocol of conversation was different. Call-waiting, a popular feature in the West, is considered impolite in some parts of Asia. This cultural blunder could have been avoided had the team captured the project environment requirements and involved the customer.

It is often the simplest things that can cause trouble since unsurprisingly in different countries people do things differently. One of the most notorious examples of this is also one of the most simple: date formats. What day and month is 2/8/2009? Of course it depends where you come from; in North America it is February 8th while in Europe (and much of the rest of the world) it is 2nd August. Clearly, when schedules and deadlines are being defined it is important that everyone is clear on the format used.

The diversity of practices and cultures and its impact on products in general and on software in particular, goes well beyond the date issue. You may be managing a project to create a new website for a company that sells products worldwide. There are language and presentation style issues to take into consideration; converting the site into different languages isn't enough. It is obvious to ensure that the translation is correct, however, the presentation layer will have its own set of requirements for different cultures. The left side of a web site may be the first focus of attention for an American; the right side would be the initial focus for anyone from the Middle East, as both Arabic and Hebrew are written from right to left. Colors also have different meanings in different cultures. White, which is a sign of purity in America (e.g., a bride's wedding dress), and thus would be a favored background color in North America, signifies death in Japan (e.g., a burial shroud). Table 6.1 summarizes different meanings of common colors.

Color	United States	China	Japan	Egypt	France
Red	Danger, stop	Happiness	Anger, danger	Death	Aristocracy
Blue	Sadness, melancholy	Heavens, clouds	Villainy	Virtue, faith, truth	Freedom, peace
Green	Novice, apprentice	Ming dynasty, heavens	Future, youth, energy	Fertility, strength	Criminality
Yellow	Cowardice	Birth, wealth	Grace, nobility	Happiness, prosperity	Temporary
White	Purity	Death, purity	Death	Joy	Neutrality

Table 6.1: The meaning of colors in various cultures Adapted from P. Russo and S. Boor, How Fluent is Your Interface? Designing for International Users, Proceedings of the INTERACT '93 and CHI '93, Association for Computing Machinery, Inc. (1993).
Table from Barron & Barron Project Management for Scientists and Engineers, http://cnx.org/content/col11120/1.4/

Project managers in multicultural projects must appreciate the culture dimensions and try to learn relevant customs, courtesies, and business protocols before taking responsibility for managing an international project. A project manager must take into consideration these various cultural influences and how they may affect the project's completion, schedule, scope and cost.

6.3 Management Knowledge and Skills

As the project manager you have to rely on your project management knowledge and your general management skills. In this area we are thinking of items like your ability to plan the project, to execute the project properly and of course to control the project and bring it to a successful conclusion with the ability to guide the project team while achieving project objectives and balancing the project constraints.

There is more to project management than just getting the work done. Inherent to the process of project management are the general management skills that allow the project manager to complete the project with some level of efficiency and control. In some respects, managing a project is similar to running a business: there are risk and rewards, finance and accounting activities, human resource issues, time management, stress management, and a purpose for the project to exist. General management skills are needed in just about every project.

6.4   Interpersonal Skills

Last but not least you also have to bring the ability onto the project to manage personal relationships as well as dealing with issues as they arise. Here were talking about your interpersonal skills as shown in Figure 6.3.

6.4.1   Communication

Project managers spend 90% of their time communicating. Therefore they must be good communicators, promoting clear unambiguous exchange of information. As a project manager, it is your job to keep a number of people well informed. It is essential that your project staff know what is expected of them: what they have to do, when they have to do it, and what budget and time constraints and quality specification they are working towards. If project staff does not know what their tasks are, or how to accomplish them, then the entire project will grind to a halt. If you do not know what the project staff is (or often is not) doing then you will be unable to monitor project progress. Finally, if you are uncertain of what the customer expects of you, then the project will not even get off the ground. Project communication can thus be summed up as who needs what information and when.

Interpersonal Skills	
Communication	Influence
Leadership	Motivation
Negotiation	Problem solving

Figure 6.3: Interpersonal skills required of a project manager.
Table from Barron & Barron Project Management for Scientists and Engineers, http://cnx.org/content/col11120/1.4/

All projects require sound communication plans, but not all projects will have the same types of communication or the same methods for distributing the information. For example, will information be distributed via mail or e-mail, is there a shared web site, or are face-to-face meetings required? The communication management plan documents how the communication needs of the stakeholders will be met, including the types of information that will be communicated, who will communicate it, who receives the communication, the methods used to communicate, the timing and frequency, the method for updating the plan as the project progresses, escalation process, and a glossary of common terms.

6.4.2   Influence

Project management is about getting things done. Every organization is different in its policies, modes of operations and underlying culture. There are political alliances, differing motivations, confecting interest, and power struggles within every organization. A project manager must understand all of the unspoken influences at work within an organization.

6.4.3 Leadership

Leadership is the ability to motivate and inspire individuals to work towards expected results. Leaders inspire vision and rally people around common goals. A good project manager can motivate and inspire the project team to see the vision and value of the project. The project manager as a leader can inspire the project team to find a solution to overcome the perceived obstacles to get the work done.

6.4.4 Motivation

Motivation helps people work more efficiently and produce better results. Motivation is a constant process that the project manager must have to help the team move towards completion with passion and a profound reason to complete the work. Motivating the team is accomplished by using a variety of team building techniques and exercises. Team building is simply getting a diverse group of people to work together in the most efficient and effective manner possible. This may involve management events as well as individual actions designed to improve team performance.

Recognition and rewards are an important part of team motivations. They are formal ways of recognizing and promoting desirable behavior and are most effective when carried out by the management team and the project manager. Consider individual preferences and cultural differences when using rewards and recognition. Some people don't like to be recognized in front of a group; others thrive on it.

6.4.5 Negotiation

Project managers must negotiate for the good of the project. In any project, the project manager, the project sponsor, and the project team will have to negotiate with stakeholders, vendors, and customers to reach a level of agreement acceptable to all parties involved in the negotiation process.

6.4.6 Problem Solving

Problem solving is the ability to understand the heart of a problem, look for a viable solution, and then make a decision to implement that solution. The premise for problem solving is problem definition. Problem definition is the ability to understand the cause and effect of the problem; this centers on root cause analysis. If a project manager treats only the symptoms of a problem rather than its cause, the symptoms will perpetuate and continue through the project life. Even worse treating a symptom may result in a greater problem. For example, increasing the ampere rating of a fuse in your car because the old one keeps blowing does not solve the problem of an electrical short that could result in a free. Root cause analysis looks beyond the immediate symptoms to the cause of the symptoms, which then affords opportunities for solutions. Once the root of a problem has been identified, a decision must be made to effectively address the problem.

Solutions can be presented from vendors, the project team, the project manager or various stakeholders. A viable solution focuses on more than just the problem; it looks at the cause and effect of the solution itself. In addition, a timely decision is needed or the window of opportunity may pass and then a new decision will be needed to address the problem. As in most cases, the worst thing you can do is nothing.

All of these interpersonal skills will be used in all areas of project management. Start practicing now because it's guaranteed that you'll need these skills on your next project.

The project manager and project team have one shared goal: to carry out the work of the project for the purpose of meeting the project's objectives. Every project has beginnings, a middle period during which activities move the project toward completion, and an ending (either successful or unsuccessful). A standard project typically has the following four major phases (each with its own agenda of tasks and issues): initiation, planning, implementation, and closure. Taken together, these phases represent the path a project takes from the beginning to its end and are generally referred to as the project life cycle .

7.1   Initiation Phase

During the first of these phases, the initiation phase, the project objective or need is identified; this can be a business problem or opportunity. An appropriate response to the need is documented in a business case with recommended solution options. A feasibility study is conducted to investigate whether each option addresses the project objective and a final recommended solution is determined. Issues of feasibility ("can we do the project?") and justification ("should we do the project?") are addressed.

Once the recommended solution is approved, a project is initiated to deliver the approved solution and a project manager is appointed. The major deliverables and the participating work groups are identified and the project team begins to take shape. Approval is then sought by the project manager to move on the detailed planning phase.

7.2   Planning Phase

The next phase, the planning phase, is where the project solution is further developed in as much detail as possible and you plan the steps necessary to meet the project's objective. In this step, the team identifies all of the work to be done. The project's tasks and resource requirements are identified, along with the strategy for producing them. This is also referred to as scope management. A project plan is created outlining the activities, tasks, dependencies and timeframes. The project manager coordinates the preparation of a project budget; by providing cost estimates for the labor, equipment and materials costs. The budget is used to monitor and control cost expenditures during project implementation.

Once the project team has identified the work, prepared the schedule and estimated the costs, the three fundamental components of the planning process are complete. This is an excellent time to identify and try to deal with anything that might pose a threat to the successful completion of the project. This is called risk management. In risk management, "high-threat" potential problems are identified along with the action that is to be taken on each high threat potential problem, either to reduce the probability that the problem will occur or to reduce the impact on the project if it does occur. This is also a good time to identify all project stakeholders, and to establish a communication plan describing the information needed and the delivery method to be used to keep the stakeholders informed.

Finally, you will want to document a quality plan; providing quality targets, assurance, and control measures along with an acceptance plan; listing the criteria to be met to gain customer acceptance. At this point, the project would have been planned in detail and is ready to be executed.

7.3   Implementation Phase

During the third phase, the implementation phase, the project plan is put into motion and performs the work of the project. It is important to maintain control and communicate as needed during implementation. Progress is continuously monitored and appropriate adjustments are made and recorded as variances from the original plan. In any project a project manager will spend most of their time in this step. During project implementation, people are carrying out the tasks and progress information is being reported through regular team meetings. The project manager uses this information to maintain control over the direction of the project by measuring the performance of the project activities comparing the results with the project plan and takes corrective action as needed. The first course of action should always be to bring the project back on course, i.e., to return it to the original plan. If that cannot happen, the team should record variations from the original plan and record and publish modifications to the plan. Throughout this step, project sponsors and other key stakeholders should be kept informed of project status according to the agreed upon frequency and format. The plan should be updated and published on a regular basis.

Status reports should always emphasize the anticipated end point in terms of cost, schedule and quality of deliverables. Each project deliverable produced should be reviewed for quality and measured against the acceptance criteria. Once all of the deliverables have been produced and the customer has accepted the final solution, the project is ready for closure.

7.4   Closing phase

During the final closure, or completion phase, the emphasis is on releasing the final deliverables to the customer, handing over project documentation to the business, terminating supplier contracts, releasing project resources and communicating the closure of the project to all stakeholders. The last remaining step is to conduct lessons learned studies; to examine what went well and what didn't. Through this type of analysis the wisdom of experience is transferred back to the project organization, which will help future project teams.

PART II – PROJECT STRATEGY

Waxing Moon Copyright © NASA
http://www.flickr.com/photos/gsfc/5837031188/

A project is successful when it achieves its objectives and meets or exceeds the expectations of the stakeholders. But who are the stakeholders? Stakeholders are individuals who either care about or have a vested interest in your project. They are the people who are actively involved with the work of the project or have something to either gain or lose as a result of the project. When you manage a project to add lanes to a highway, motorists are stakeholders who are positively affected. However, you negatively affect residents who live near the highway during your project (with construction noise) and after your project with far reaching implications (increased traffic noise and pollution).

NOTE: Key stakeholders can make or break the success of a project. Even if all the deliverables are met and the objectives are satisfied, if your key stakeholders aren't happy, nobody's happy.

The project sponsor, generally an executive in the organization with the authority to assign resources and enforce decisions regarding the project, is a stakeholder. The customer, subcontractors, suppliers and sometimes even the Government are stakeholders. The project manager, project team members and the managers from other departments in the organization are stakeholders as well. It's important to identify all the stakeholders in your project upfront. Leaving out important stakeholders or their department's function and not discovering the error until well into the project could be a project killer.

Figure 8.1 shows a sample of the project environment featuring the different kinds of stakeholders involved on a typical project. A study of this diagram confronts us with a couple of interesting facts.

First, the number of stakeholders that project managers must deal with assures that they will have a complex job guiding their project through the lifecycle. Problems with any of these members can derail the project.

The diagram also shows that project managers have to deal with people external to the organization as well as the internal environment, certainly more complex than what a manager in an internal environment faces. For example, suppliers who are late in delivering crucial parts may blow the project schedule. To compound the problem, project managers generally have little or no direct control over any of these individuals.

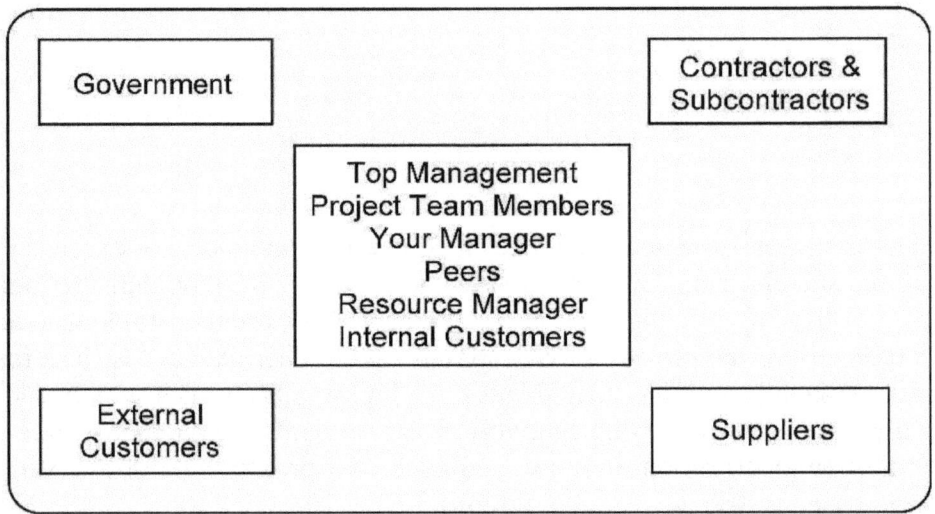

Figure 8.1: Project stakeholders.

Let's take a look at these stakeholders and their relationships to the project manager.

8.1   Top Management

Top management may include the president of the company, vice presidents, directors, division managers, the corporate operating committee, and others. These people direct the strategy and development of the organization.

On the plus side, you are more likely to have top management support, which means it will be easier to recruit the best staff to carry out the project, and to acquire needed material and resources; also visibility can enhance a PM's professional standing in the company.

On the minus side, failure can be quite dramatic and visible to all, and if the project is large and expensive (most are), the cost of failure will be more substantial than for a smaller less visible project.

Some suggestions in dealing with top management are:

- Develop in-depth plans and major milestones that must be approved by top management during the planning and design phases of the project.

- Ask top management associated with your project for their information reporting needs and frequency.

- Develop a status reporting methodology to be distributed on a scheduled basis.

- Keep them informed of project risks and potential impacts at all times.

8.2 The Project Team

The project team is those people dedicated to the project or borrowed on a part-time basis. As project manager you need to provide leadership, direction, and above all, the support to team members as they go about accomplishing their tasks. Working closely with the team to solve problems can help you learn from the team and build rapport. Showing your support for the project team and for each member will help you get their support and cooperation.

Some difficulties in dealing with project team members include:

- Because project team members are borrowed and they don't report to you, their priorities may be elsewhere.

- They may be juggling many projects as well as their full time job and have difficulty meeting any deadline.

- Personality conflicts may arise. These may be caused by differences in social style or values or they may be the result of some bad experience when people worked together in the past.

- You may find out about missed deadlines when it is too late to recover.

Managing project team members requires interpersonal skills. Here are some suggestions that can help:

- Involve team members in project planning.

- Arrange to meet privately and informally with each team member at several points in the project, perhaps for lunch or coffee.

- Be available to hear team members' concerns at any time.

- Encourage team members to pitch in and help others when needed.

- Complete a project performance review for team members.

8.3 Your Manager

Typically the boss decides what our assignment is and who can work with us on our projects. Keeping your manager informed will help ensure that you get the necessary resources to complete your project.

- If things go wrong on a project, it is nice to have an understanding and supportive boss to go to bat for us if necessary. By supporting your manager, you will find your manager will support you more often.

- Find out exactly how your performance will be measured.

- When unclear about directions, ask for clarification.

- Develop a reporting schedule that is acceptable to your boss.

- Communicate frequently.

8.4 Peers

Peers are people on the project team or not, who are at the same level in the organization as you. These people will, in fact, also have a vested interest in the product. However, they will have neither the leadership responsibilities nor the accountability for the success or failure of the project that you have.

Your relationship with peers can be impeded by:

- Inadequate control over peers.

- Political maneuvering or sabotage.

- Personality conflicts or technical conflicts.

- Envy because your peer may have wanted to lead the project.

- Conflicting instructions from your manager and your peer's manager.

Peer support is essential. Because most of us serve our self-interest first, use some investigating, selling, influencing and politicking skills here. To ensure you have cooperation and support from your peers:

Get the support of your project sponsor or top management to empower you as the project manager with as much authority as possible. It's important that the sponsor makes it clear to the other team members that their cooperation on project activities is expected.

- Confront your peer if you notice a behavior that seems dysfunctional, such as bad-mouthing the project.

- Be explicit in asking for full support from your peers. Arrange for frequent review meetings.

- Establish goals and standards of performance for all team members.

8.5 Resource Managers

Because project managers are in the position of borrowing resources, other managers control their resources. So their relationships with people are especially important. If their relationship is good, they may be able to consistently acquire the best staff and the best equipment for their projects. If relations aren't so good, they may find themselves not able to get good people or equipment needed on the project.

8.6 Internal Customer

Internal customers are individuals within the organization who have projects that meet the needs of internal demands. The customer holds the power to accept or reject your work. Early in the relationship, the project manager will need to negotiate, clarify, and document project specifications and deliverables. After the project begins, the project manager must stay tuned in to the customer's concerns and issues and keep the customer informed.

Common stumbling blocks when dealing with internal customers include:

- A lack of clarity about precisely what is wanted by the customer.

- A lack of documentation for what is wanted.

- A lack of knowledge of the customer's organization and operating characteristics.

- Unrealistic deadlines, budgets, or specifications.

- Hesitancy to sign off on the project or accept responsibility for decisions.

- Changes in project scope.

To meet the needs of the customer, client or owner, be sure to do the following:

- Learn the client organization's buzzwords, culture, and business.

- Clarify all project requirements and specifications in a written agreement.

- Specify a change procedure.

- Establish the project manager as the focal point of communications in the project organization.

8.7 External customer

External customers are the customers when projects could be marketed to outside customers. In the case of Ford Motor Company for example, the external customers would be the buyers of the automobiles. Also if you are managing a project at your company for Ford Motor Company, they will be your external customer.

8.8 Government

Project managers working in certain heavily regulated environment (e.g., pharmaceutical, banking, military industries, etc.) will have to deal with government regulators and departments. These can include all or some levels from city, through county, state, and federal, to international.

8.9 Contractors, subcontractors, and suppliers

There are times when organizations don't have the expertise in-house or available resources, and work is farmed out to contractors or subcontractors. This can be a construction management foreman, network consultant, electrician, carpenter, architect, and in general anyone who is not an employee. Managing contractors or suppliers requires many of the skills needed to manage full-time project team members.

Any number of problems can arise with contractors or subcontractors:

- Quality of the work.

- Cost overruns

- Schedule slippage

Many projects depend on goods provided by outside suppliers. This is true for example of construction projects where lumber, nails, brick and mortar come from outside suppliers. If the supplied goods are delivered late or in short supply or of poor quality or if the price is greater than originally quoted, the project may suffer.

Depending on the project, managing relationships can consume more than half of the project manager's time. It is not purely intuitive; it involves a sophisticated skill set that includes managing conflicts, negotiating, and other interpersonal skills.

Many times, project stakeholders have conflicting interests. It's the project manager's responsibility to understand these conflicts and try to resolve them. It's also the project manger's responsibility to manage stakeholder expectations. Be certain to identify and meet with all key stakeholders early in the project to understand all their needs and constraints.

Project managers are somewhat like politicians. Typically, they are not inherently powerful, or capable of imposing their will directly to co-workers, subcontractors and suppliers. Like politicians, if they are to get their way, they have to exercise influence effectively over others. On projects, project managers have direct control over very few things; therefore their ability to influence others - to be a good politician - may be very important

Here are a few steps a good project politician should follow. However, a good rule is that when in doubt, stakeholder conflicts should always be resolved in favor of the customer.

9.1 Assess the environment

Identify all the relevant stakeholders. Because any of these stakeholders could derail the project, we need to consider their particular interest in the project.

- Once all relevant stakeholders are identified, we try to determine where the power lies.

- In the vast cast of characters we confront, who counts most?

- Whose actions will have the greatest impact?

9.2 Identify goals

After determining who the stakeholders are, we should identify their goals.

- What is it that drives them?

- What is each after?

- We should also be aware of hidden agendas or goals that are not openly articulated.

- We need to pay special attention to the goals of the stakeholders who hold the power.

9.3 Define the problem

- The facts that constitute the problem should be isolated and closely examined.

- The question "What is the real situation?" should be raised over and over.

The project initiation phase is the first phase within the project management life cycle, as it involves starting up a new project. Within the initiation phase, the business problem or opportunity is identified, a solution is defined, a project is formed, and a project team is appointed to build and deliver the solution to the customer. A business case is created to define the problem or opportunity in detail and identify a preferred solution for implementation. The business case includes:

- A detailed description of the problem or opportunity

- A list of the alternative solutions available

- An analysis of the business benefits, costs, risks and issues

- A description of the preferred solution

- A summarized plan for implementation

The project sponsor then approves the business case, and the required funding is allocated to proceed with a feasibility study. It is up to the project sponsor to determine if the project is worth undertaking and whether the project will be profitable to the organization. The completion and approval of the feasibility study triggers the beginning of the planning phase. The feasibility study may also show that the project is not worth pursuing and the project is terminated; thus the next phase never begins.

All projects are created for a reason. Someone identifies a need or an opportunity and devises a project to address that need. How well the project ultimately addresses that need defines the project's success or failure.

The success of your project depends on the clarity and accuracy of your business case and whether people believe they can achieve it. Whenever you consider past experience, your business case is more realistic; and whenever you involve people in the business case's development, you encourage their commitment to achieving it.

Often the pressure to get results encourages people to go right into identifying possible solutions without fully understanding the need; what the project is trying to accomplish. This strategy can create a lot of immediate activity but it also creates significant chances for waste and mistakes if the wrong need is addressed. One of the best ways to gain approval for a project is to clearly identify the project's objectives and describe the need or opportunity for which the project will provide a solution.

For most of us, being misunderstood is a common occurrence, something that happens on a daily basis. At the restaurant the waiter brings us our dinner and we note that the baked potato is filled with sour cream, even though we expressly requested "no sour cream". Projects are filled with misunderstandings between customers and project staff. What the customer orders (or more accurately what they think they order) is often not what they get. The cliché is "I know that's what I said, but it's not what I meant" The cartoon demonstrates the importance of establishing clear objectives.

The need for establishing clear project objectives cannot be overstated. An objective or goal lacks clarity if, when shown to five people, it is interpreted in multiple ways. Ideally, if an objective is clear, you can show it to five people who, after reviewing it, hold a single view about its meaning. The best way to make an objective clear is to state it such a way that it can be verified. Building in measures can do this. It is important to provide quantifiable definitions to qualitative terms.

For example, an objective of the team principle (project manager) of a Formula 1 racing team may be that their star driver, "finish the lap as fast as possible." That objective is filled with ambiguity.

How fast is "fast as possible?" Does that mean the fastest lap time (the time to complete one lap) or does it mean the fastest speed as the car crosses the start/finish line (that is at the finish of the lap)?

By when should the driver be able to achieve the objective? It is no use having the fastest lap after the race has finished, and equally the fastest lap does not count for qualifying, and therefore starting position, if it is performed during a practice session.

The ambiguity of this objective can be seen from the following example. Ferrari's Michael Schumacher achieved the race lap record at the Circuit de Monaco of 1 min 14.439 sec in 2004 (Figure 10.2). However, he achieved this on lap 23 of the race, but crashed on lap 45 of a 77 lap race So while he achieved a fastest lap and therefore met the specific project goal of "finish the lap as fast as possible", it did not result in winning the race, clearly a different project goal. In contrast, the fastest qualifying time at the same event was by Renault's Jarno Trulli (1 min 13.985 sec), which gained him pole position for the race, in which he went on to win (Figure 10.3). In his case he also achieved the specific project goal of "finish the lap as fast as possible", but also the larger goal of winning the race.

Figure 10.2: Despite achieving the project goal of the "finish the lap as fast as possible" Ferrari's Michael Schumacher crashed 21 laps later and did not finish the race.
Photo from Barron & Barron Project Management for Scientists and Engineers, http://cnx.org/content/col11120/1.4/

Figure 10.3: Renault's Jarno Trulli celebrating his win at the 2004 Monaco Grand Prix.

The objective can be strengthened considerably if it is stated as follows: "To be able to finish the 3.340 km lap at the Circuit de Monaco at the Monaco Grand Prix in 1 min 14.902 sec or less, during qualifying on 23th May, 2009. This was the project objective achieved by Brawn GP's Jenson Button (Figure 10.4).

Figure 10.4..: Jenson Button took his Brawn GP car to pole position at the Monaco Grand Prix with a lap time of 1 min 14.902 sec. He also went on to with the race, even though he did not achieve that lap time during the race.

There is still some ambiguity in this objective; for example, it assumes the star driver will be driving the team's race car and not a rental car from Hertz. However, it clarifies the team principal's intent quite nicely. It should be noted that a clear goal is not enough. It must also be achievable. The team principal's goal becomes unachievable, for example, if he changes it to require his star driver finish the 3.340 km lap in 30 seconds or less.

To ensure the project's objectives are achievable and realistic, they must be determined jointly by managers and those who perform the work. Realism is introduced because the people who will do the work have a good sense of what it takes to accomplish a particular task. In addition, this process assures some level of commitment on all sides; management expresses its commitment to support the work effort and workers demonstrate their wiliness to do the work.

Imagine you are an office manager and you have contracted a painter to paint your office. Your goal or objective is to have the office painted a pleasing blue color. Consider the following conversation that occurs after the job was finished:

Figure 10.5: The consequence of not making your objective clear.
Illustration from Barron & Barron Project Management for Scientists and Engineers, http://cnx.org/content/col11120/1.4/

This conversation captures in a nutshell the essence of a major source of misunderstandings on projects: The importance of setting clear objectives. The office manager's description of how he wanted the room painted meant one thing to him and another to the painter. As a consequence, the room was not painted to the office manager's satisfaction. Had his objective been more clearly defined, he probably would have had what he wanted.

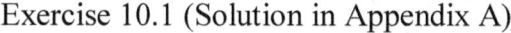

Exercise 10.1 (Solution in Appendix A)

How could you make the objective for painting a room clear such that the office manager gets what he wanted?

Certification in project management is available from the Project Management Institute, PRINCE2, ITIL, Critical Chain, and others. Agile project management methodologies (Scrum, Extreme Programming, Lean Six Sigma, others) also have certifications.

11.1   Project Management Institute Overview

Five volunteers founded the Project Management Institute (PMI) in 1969. Their initial goal was to establish an organization where members could share their experiences in project management and to discuss issues. Today, PMI is a non-profit project management professional association and the most widely recognized organization in terms of promoting project management best practices. PMI was formed to serve the interests of the project management industry. The premise of PMI is that the tools and techniques of project management are common even among the widespread application of projects from the software to the construction industry. PMI first began offering the PMP certification exam in 1984. Although it took a while for people to take notice, now more than 260,000 individuals around the world hold the PMP designation.

To help keep project management terms and concepts clear and consistent, PMI introduced the book Project Management Body of Knowledge (PMBOK) Guide in 1987. It was updated it in 1996, 2000, 2004, and most recently in 2009 as the fourth edition. At present, there are more than 1 million copies of the PMBOK Guide in circulation. The highly regarded Institute of Electrical and Electronics Engineers (IEEE) has adopted it as their project management standard. In 1999 PMI was accredited as an American National Standards Institute (ANSI) standards developer and also has the distinction of being the first organization to have its certification program attain International Organization for Standardization (ISO) 9001 recognition. In 2008, the organization reported more than 260,000 members in over 171 countries. PMI has its headquarters in Pennsylvania, USA and also has offices in Washington, D.C., and Canada, Mexico, Beijing, China, as well as Regional Service Centers in Singapore, Brussels (Belgium) and New Delhi (India). Recently, an office was opened in Mumbai (India).

11.2   Scrum Development Overview

Scrum is another formal project management/product development methodology and part of agile project management. Scrum is a term from rugby (scrummage) that means a way of restarting a game. It's like restarting the project efforts every X weeks. It's based on the idea that you do not really know how to plan the whole project up front, so you start and build empirical data, and then re-plan and iterate from there.

Scrum uses sequential Sprints for development. Sprints are like small project phases (ideally 2 to 4 weeks). The idea is to take one day to plan for what can be done now, then develop what was planned for, and demonstrate it at the end of the Sprint. Scrum uses a short daily meeting of the development team to check what was done yesterday, what is planned for the next day, and what if anything is impeding the team members from accomplishing what they have committed to. At the end of the Sprint, what has been demonstrated can then be tested, and the next Sprint cycle starts.

Scrum methodology defines several major roles. They are:

- Product Owner(s): essentially the business owner of the project who knows the industry, the market, the customers and the business goals of the project. The Product Owner **must** be intimately involved with the Scrum process, especially the planning and the demonstration parts of the Sprint.

- Scrum Master: somewhat like a project manager, but not exactly. The Scrum Master's duties are essentially to: remove barriers that impede the progress of the development team, teach the Product Owner how to maximize ROI in terms of development effort, facilitate creativity and empowerment of team, improve the productivity of the team, improve engineering practices and tools, run daily standup meetings, track progress, and ensure the health of the team.

- Development Team: self organizing (light touch leadership), empowered, participate in planning and estimating for each Sprint, do the development, and demonstrate the results at the end of the Sprint. It has been shown that the ideal size for a development team is 7 +/- 2. The development team can be broken into teamlets that _swarm' on user stories, which are created in the Sprint planning session.

- Typically, the way a product is developed is that there is a Front Burner (which has stories/tasks for the current Sprint), a Back Burner (which has stories for the next Sprint), and a Fridge (which has stories for later, as well as process changes). One can look at a Product as having been broken down like this: Product -> Features -> Stories -> Tasks

Often effort estimations are done using _Story Points' (Tiny = 1 SP, Small = 2 SP, Medium = 4 SP, Large = 8 SP, Big = 16+ SP, Unknown = ? SP) Stories can be of various types. User stories are very common and are descriptions of what the user can do and what happens as a result of different actions from a given starting point. Other types of stories are: Analysis, Development, QA, Documentation, Installation, Localization, Training, etc.

Planning meetings for each Sprint require participation by the Product Owner, the Scrum Master, and the Development Team. In the planning meeting, they set the goals for the upcoming Sprint and select a subset of the product backlog (proposed stories) to work on. The Development team de-composes stories to tasks and estimates them, and the Development team and Product Owner do final negotiations to determine the backlog for the following Sprint.

Scrum uses metrics to help with future planning and tracking of progress. A few of them are: Burn down – The number of hours remaining in the Sprint versus the time in days. Velocity – Essentially, how much effort the team completes per Sprint. (After approximately 3 Sprints with the same team, one can get a feel for what the team can do going forward.)

Some Caveats about using Scrum methodology: 1) You need committed, mature developers, 2) You still need to do major requirements definition, some analysis, architecture definition, and definition of roles and terms up-front or early, 3) You need commitment from company and the Product Owner, and 4) It is best for products that require frequent new releases or updates, and less good for large, totally new products that will not allow for frequent upgrades once they are released.

Chapter 12: Culture and Project Management

12.1   What is Organizational Culture?

When working with internal and external customers on a project, it is essential to pay close attention to relationships, context, history and the organizational culture. Corporate culture refers to the beliefs, attitudes, and values that the organization's members share and to the behaviors consistent with them (that they give rise to). Corporate culture sets one organization apart from another, and dictates how members of the organization will see you, interact with you, and sometimes judge you. Often, projects too have a specific culture, work norms and social conventions.

Some aspects of corporate culture are easily observed; others are more difficult to discern. You can easily observe the office environment and how people dress and speak. In one company individuals work separately in closed offices; in others, teams may work in shared environments. The more subtle components of corporate culture, such as the values and overarching business philosophy, may not be readily apparent, but they are reflected in member behaviors, symbols and conventions used.

12.2   Project Manager's Checklist:

Once the corporate culture has been identified, members should try to adapt to the frequency, formality, and type of communication customary in that culture. This adaptation will strongly affect project members' productivity and satisfaction internally, as well as with the client-organization.

- Which stakeholders will make the decision in this organization on this issue? Will your project decisions and documentation have to go up through several layers to get approval? If so, what are the criteria and values that may affect acceptance there? For example, is being on schedule the most important consideration? Cost? Quality?

- What type of communication among and between stakeholders is preferred? Do they want lengthy documents? Is ―short and sweet" the typical standard?

- What medium of communication is preferred? What kind of medium is usually chosen for this type of situation? Check the files to see what others have done. Ask others in the organization.

- What vocabulary and format are used? What colors and designs are used? (i.e., at Hewlett-Packard (HP), all rectangles have curved corners)

12.3 Project Team Challenges

Today's globally-distributed organizations (and projects) consist of people who have a different ―worldview". Worldview is a looking glass through which [people] see the world as quoted by Bob Shebib (Shebib, 2003. p. 296): "[It is] a belief system about the nature of the universe, its perceived effect on human behavior, and one's place in the universe. Worldview is a fundamental core set of assumptions explaining cultural forces, the nature of humankind, the nature of good and evil, luck, fate, spirits, the power of significant others, the role of time, and the nature of our physical and natural resources."

If, for example, a US manager is sent to India to manage an R&D team or a joint-venture, s/he is likely to have to —[cope] with eco-shock or the physiological, psychological, and social reaction to a new assignment ecology". Hanging one's shingle in a fluid and culturally-diverse organization, project team and work culture; new working relationships and hidden challenges have significant implications for performance and knowledge exchange – for the manager and his/her colleagues at home and in the host country.

In most situations there is simply **no** substitute for having a well-placed person from the host culture to guide the new person through the cultural nuances of getting things done. In fact, if this 'intervention' isn't present, it is likely to affect the person's motivation or desire to continue trying to break through the cultural (and other) barriers. Indeed, optimal effectiveness in such situations requires learning of developing third-world cultures or international micro cultures, shared perceptions among the culturally diverse task participants on how to get things done. Project leaders require sensitivity and awareness of multicultural preferences. The following broad areas should be considered:

- Individual identity and role within project vs. family-of-origin and community

- Verbal and emotional expressiveness

- Relationship expectations

- Style of communication

- Language

- Personal priorities, values and beliefs

- Time Orientation

There are many interpersonal dynamics and intra-project challenges faced by a globally-distributed team. Individual members and the team itself requires important social supports to mitigate uncertainty, conflict, motivational challenges, culture shock and the more-encompassing eco-shock– that comes from facing head-on the unfamiliar and diverse situations consistent with a different cultural and distributed context.

Diverse and globally distributed project teams (i.e., different ethnic cultures, genders, age, and functional capabilities) often working on complex projects spanning multiple time zones, geography and history, operating with tight deadlines in cost-conscious organizations, need to make time and resources available to physically meet each other, and connect (at the very least) at a formal _kick-off* meeting. Especially when working with team members from high-context cultures it is essential to meet face-to-face, and discover member's individual identities, cultural preferences and share professional knowledge and personal stories; observe critical verbal and non-verbal cues (that may not easily be observed online, or on the telephone).

This is key to establishing a safer climate and building trust for stronger relationships among both team members and management.

12.4   Dealing with Conflict

The question isn't whether, when or what will create conflict among intercultural team members — or with what frequency it will occur. If a team wants to overcome (or harness) conflict for effectiveness and productivity, the question is how to navigate and resolve the conflicts. Conflict that springs from diversity can actually assist the team in completing complex problem-solving. However, if not navigated successfully, it can create relationship strain and derail achievement due to increased difficulties in communication and coordination.

As the global marketplace continues its rapid expansion, researchers are increasingly turning their attention to the issue of conflict management. Differing social and cultural values don't necessarily increase the number of conflicts a team will experience, but they can have an impact on how conflicts get managed and resolved. Cultural awareness is needed for understanding and appreciating others' values and behavioral norms. Without that, Global Holdings' foreign assignments will become an overwhelming challenge. Self-awareness and skill development can aid in resolving the problematic conflict arising from cultural differences to help the team maintain good relations and remain productive.

12.4   Bibliography for Chapter 12

See Appendix C for references.

PART III – PROJECT PLANNING

Full Moon Copyright © Chris Ptacek
http://www.flickr.com/photos/chrisptacek/5541729614/sizes/z/in/photostream/

After the project has been defined and the project team has been appointed, you are ready to enter the second phase in the project management life cycle: the detailed project planning phase.

Project planning is at the heart of the project life cycle, and tells everyone involved to where you're going and how you're going to get there. The planning phase is when the project plans are documented, the project deliverables and requirements are defined, and the project schedule is created. It involves creating a set of plans to help guide your team through the implementation and closure phases of the project. The plans created during this phase will help you managing time, cost, quality, changes, risk and related issues. It will also help you controlling staff and external suppliers, to ensure that you deliver the project on time, budget and within schedule.

The project planning phase is often the most challenging phase for a project manager, as you need to make an educated guess about the staff, resources and equipment needed to complete your project. You may also need to plan your communications and procurement activities, as well as contract any 3rd party suppliers.

The purpose of the project planning phase is as follows:

- Establishing business requirements.

- Establishing cost, schedule, list of deliverables, and delivery dates.

- Establishing resources plan.

- Getting management approval and proceeding to the next phase.

The basic processes of the project planning are:

- Scope planning – specifying the in-scope requirements for the project and facilitates creating the work breakdown structure.

- Preparing the Work Breakdown Structure – spelling out the breakdown of the project into tasks and sub tasks.

- Project schedule development – listing the entire schedule of the activities and detailing their sequence of implementation.

- Resource planning – indicating who will do what work, at which time and if any special skills are needed to accomplish the project tasks.

- Budget planning – specifying the budgeted cost to be incurred at the completion of the project.

- Procurement planning – focusing on vendors outside your company, sub-contracting.

- Risk management – planning for possible risks, considering optional contingency plans and mitigation strategies.

- Quality planning – assessing quality criteria to be used for the project.

- Communication planning – designing the communication strategy with all project stakeholders.

The planning phase refines the project's objectives, which were gathered during the initiation phase. It includes planning the steps necessary to meet those objectives, by further identifying the specific activities and resources required to complete the project. Now that these objectives have been recognized, they must be clearly articulated, detailing an in-depth scrutiny of each recognized objective. With such scrutiny, our understanding of the objective may change. Often the very act of trying to describe something precisely, gives us a better understanding of what we are looking at. This articulation serves as the basis for the development of requirements. What this means is that after an objective has been clearly articulated, we can describe it in concrete (measurable) terms, what we have to do to achieve it. Obviously, if we do a poor job of articulating the objective, our requirements will be misdirected and the resulting project will not represent the true need.

Users will often begin describing their objectives in qualitative language. The project manager must work with the user to provide quantifiable definitions to those qualitative terms. These quantifiable criteria include: schedule, cost, and quality measures. In the case of project objectives, these elements are used as measurements to determine project satisfaction and successful completion. Subjective evaluations are replaced by actual numeric attributes.

Example 13.1

A web user may ask for a fast system. The quantitative requirement should be all screens must load in under 3 seconds. Describing the time limit during which the screen must load is specific and tangible. For that reason, you'll know that the requirement has been successfully completed when the objective has been met.

Example 13.2

Let's say that your company is going to produce a holiday batch of eggnog. Your objective statement might be stated this way: Christmas Cheer, Inc. will produce two million cases of holiday eggnog, to be shipped to our distributors by October 30, at a total cost of $1.5 million or less. The objective criteria in this statement are clearly stated and successful fulfillment can easily be measured. Stakeholders will know that the objectives are met, when the two million cases are produced and shipped by the due date within the budget stated.

When articulating the project objectives you should follow the SMART rule:

- Specific – get into the details. Objectives should be specific and written in clear, concise, and understandable terms.

- Measurable – use quantitative language so you know when you successfully complete the task. Acceptable – agreed with the stakeholders.

- Realistic – in terms of achievement. Objectives that are impossible to accomplish, are not realistic and not attainable. Objectives must be centered in reality.

- Time bound – deadlines not durations. Objectives should have a timeframe with an end date assigned to them.

If you follow these principles, you'll be certain that your objectives meet the quantifiable criteria needed to measure success.

You always want to know exactly what work has to be done **before** you start it. You've got a collection of team members, and you need to know exactly what they're going to do, in order to meet the project's objectives. The scope planning process is the very first thing you do to manage your scope. Project scope planning is concerned with the definition of all the work needed to successfully meet the project objectives. The whole idea here is that when you start the project, you need to have a clear picture of all the work that needs to happen on your project, and as the project progresses, you need to keep that scope up to date and written down in the project's scope management plan.

14.1   Defining the Scope

You already got a head start on refining the project's objectives in quantifiable terms, but now you need to plan further and write down all the intermediate and final deliverables that you and your team are going to produce over the course of the project. Deliverables include everything that you and your team produce for the project; anything that your project will deliver. The deliverables for your project include all of the products or services that you and your team are performing for the client, customer, or sponsor. They include every intermediate document, plan, schedule, budget, blueprint, and anything else that will be made along the way, including all of the project management documents you put together. Project deliverables are tangible outcomes, measurable results, or specific items that must be produced to consider either the project or the project phase completed. Intermediate deliverables like the objectives must be specific and verifiable.

All deliverables must be described in sufficiently low level of details, so that they can be differentiated from related deliverables. For example:

- A twin engine plane versus a single engine plane.

- A red marker versus a green marker.

- A daily report versus a weekly report.

- A departmental solution versus an enterprise solution.

One of the project manager's primary functions is to accurately document the deliverables of the project and then manage the project so that they are produced according to the agreed upon criteria. Deliverables are the output of each development phase, described in a quantifiable way.

14.2   Project Requirements

After all the deliverables are identified, the project manager needs to document all the requirements of the project. Requirements describe the characteristics of the final deliverable, either a product or a service. They describe the required functionality that the final deliverable must have or specific conditions the final deliverable must meet in order to satisfy the objectives of the project. A requirement is an objective that must be met. The project's requirements, defined in the scope plan, describe what a project is supposed to accomplish and how the project is supposed to be created and implemented. Requirements answer the following questions regarding the **as-is** and **to-be** states of the business: who, what where, when, how much, how does a business process work.

Requirements may include attributes like dimensions, ease of use, color, specific ingredients, and so on. If we go back to the example of the company producing holiday eggnog; one of the major deliverables is the cartons that hold the eggnog. The requirements for that deliverable may include carton design, photographs that will appear on the carton, color choices, etc.

Requirements specify what final the project deliverable should look like and what it should do. They can be divided into six basic categories, functional, non-functional, technical, user, business, and regulatory requirements.

14.3   Functional Requirements

Functional requirements describe the characteristics of the final deliverable, what emerges from the project in ordinary non-technical language. They should be understandable to the customers, and the customers should play a direct role in their development. Functional requirements are what you want the deliverable to do.

Example 14.3

If you were buying vehicles for a business, your functional requirement might be: "the vehicles should be able to take up to one ton load from a warehouse to a shop".

Example 14.4

For a computer system you may define what the system is to do: "the system should store all details of a customer's order".

The important point to note is that **what** is wanted is specified, and **not how** it will be delivered.

14.4   Non-Functional Requirements

Non-functional requirements specify criteria that can be used to judge the final product or service that your project delivers. They are restrictions or constraints to be placed on the deliverable and how to build it. Their purpose is to restrict the number of solutions that will meet a set of requirements. Using the vehicle example (Example 14.3); without any constraints, the functional requirement of a vehicle to take a load from a warehouse to a shop, the solutions being offered might result in anything from a small to a large truck. Non-functional requirements can be split into two types: performance and development.

To restrict the types of solutions you might include these performance constraints:

- The purchased trucks should be American made trucks due to government incentives. The load area must be covered.

- The load area must have a height of at least 10 feet.

Similarly, for the computer system example (Example 14.4), you might specify values for the generic types of performance constraints:

- The response time for information is displayed on the screen for the user.

- The number of hours a system should be available.

- The number of records a system should be able to hold.

- The capacity for growth of the system.

- The length of time a record should be held for auditing purposes.

For the customer records example these might be:

- The system should be available from 9 AM to 5 PM Monday to Friday.

- The system should be able to hold 100,000 customer records initially. The system should be able to add 10,000 records a year for 10 years. A record should be fully available on the system for at least 7 years. One important point with these examples is that they restrict the number of solution options that are offered to you by the developer. In addition to the performance constraints you may include some development constraints.

There are three general types of non-functional development constraints:

- Time: When a deliverable should be delivered

- Resource: How much money is available to develop the deliverable

- Quality: Any standards that are used to develop the deliverable, and development methods, etc.

14.5 Technical Requirements

Technical requirements emerge from the functional requirements, they answer the questions: how will the problem be solved this time and will it be solved technologically and/or procedurally. They answer how the system needs to be designed and implemented, to provide required functionality and fulfill required operational characteristics.

For example, in a software project, the functional requirements may stipulate that a database system will be developed to allow access to financial data through a remote terminal; the corresponding technical requirements would spell out the required data elements, the language in which the database management system will be written (due to existing knowledge in-house), the hardware on which the system will run (due to existing infrastructure), telecommunication protocols that should be used and so forth.

14.6   User Requirements

User requirements describe what the users need to do with the system or product. They focus is on the user experience with the system under all scenarios. These requirements are the input for the next development phases: user-interface design and system test cases design.

14.7   Business Requirements

Business requirements are the needs of the sponsoring organization, always from a management perspective. Business requirements are statements of the business rationale for the project. They are usually expressed in broad outcomes, satisfying the business needs, rather than specific functions the system may perform. These requirements grow out of the vision for the product that, in turn, is driven by mission (or business) goals and objectives.

14.8   Regulatory requirements

Regulatory requirements can be internal or external and are usually **non-negotiable**. They are the restrictions, licenses and laws, applicable to a product or business, imposed by the government.

14.9   An Example of Requirements

Automated teller machines (ATMs) can be used to illustrate a wide range of requirements (Figure 14.2). What are some of the physical features of these machines, and what kinds of functions do they perform for the bank's customers? Why did banks put these systems in place? What are the high level business requirements?

Figure 14.2: A typical exterior automated teller machines (ATMs).
Photo from Barron & Barron Project Management for Scientists and Engineers, http://cnx.org/content/col11120/1.4/

The following represents one possible example of each type of requirement as they would be applied to a bank's external ATM.

- ATM functional requirement: The system shall enable the user to select whether or not to produce a hardcopy transaction receipt, before completing a transaction.

- ATM non-functional requirement: All displays shall be in white, 14 pt Arial text on black background.

- ATM technical requirement: The ATM system will connect seamlessly to the existing customers' database.

- ATM user requirement: The system shall complete a standard withdrawal from a personal account, from login to cash, in less than two minutes -.

- ATM business requirement: By providing superior service to our retail customers, Monumental Bank's ATM network will allow us to increase associated service fee revenue by 10% annually on an ongoing basis, using a baseline of December 2008.

- ATM regulatory requirement: All ATMs shall connect to standard utility power sources within their civic jurisdiction, and be supplied with uninterruptible power source approved by the company.

The effective specification of requirements is one of the most challenging undertakings project managers face. Inadequately specified requirements will guarantee poor project results.

Documenting requirements is much more than just the process of writing down the requirements as the user sees them; it should cover not only what decisions have been made, but why they have been made, as well. Understanding the reasoning that was used to arrive at a decision is critical in avoiding repetition. For example, the fact that a particular feature has been excluded, because it is simply not feasible, needs to be recorded. If it is not, then the project risks wasted work and repetition, when a stakeholder requests the feature be reinstated during development or testing.

Once the requirements are documented, have the stakeholders sign off on their requirements as a confirmation of what they desire.

While the project manager is responsible for making certain the requirements are documented, it does not mean that the project manager performs this task. The project manager enlists the help of all the stakeholders: business analysts, requirement analysts, business process owners, customers and other team members to conduct the discussions, brain-storming, interviews, documenting and signing-off the requirements. The project manager is responsible only for enabling the process and facilitating it. If the project manager feels that the quality of the document is questionable, his/her duty is to stop the development process.

The project manager reviews the requirements, incorporates them into the project documentation library, and uses them as an input for the project plan.

15.1   Preparing the work breakdown structure

Now that we have the deliverables and requirements well defined, the process of breaking down the work of the project via a work breakdown structure begins. The Work Breakdown Structure (WBS) defines the scope of the project. It breaks the work down into smaller components that can be scheduled, estimated, easily monitored and controlled. The idea behind the work breakdown schedule is simple. You subdivide a complicated task into smaller tasks, until you reach a level that cannot be further subdivided.

Anyone familiar with the arrangements of folders and files in a computer memory, or who has researched their ancestral family tree, should be familiar with this idea. You stop breaking down the work when you reach a low enough level to perform a relative accurate estimate. Usually,   we estimate more accurately how long a small task will take and how much it will cost to perform, and we underestimate significantly large efforts. Each descending level of the WBS represents an increased level of detailed definition of the project work.

As an example, if I want to clean a room, I might begin by picking up clothes, toys, and other things that have been dropped on the floor. I could use a vacuum cleaner to get dirt out of the carpet. I might take down the curtains and take them to the cleaners, then dust the furniture. All of these tasks are subtasks performed to clean the room. As for vacuuming the room, I might have to get the vacuum cleaner out of the closet, connect the hose, empty the bag, and put the machine back in the closet. These are smaller tasks to be performed while accomplishing the subtask called vacuuming. The diagram in Figure 15.1 shows how this might be portrayed in WBS format.

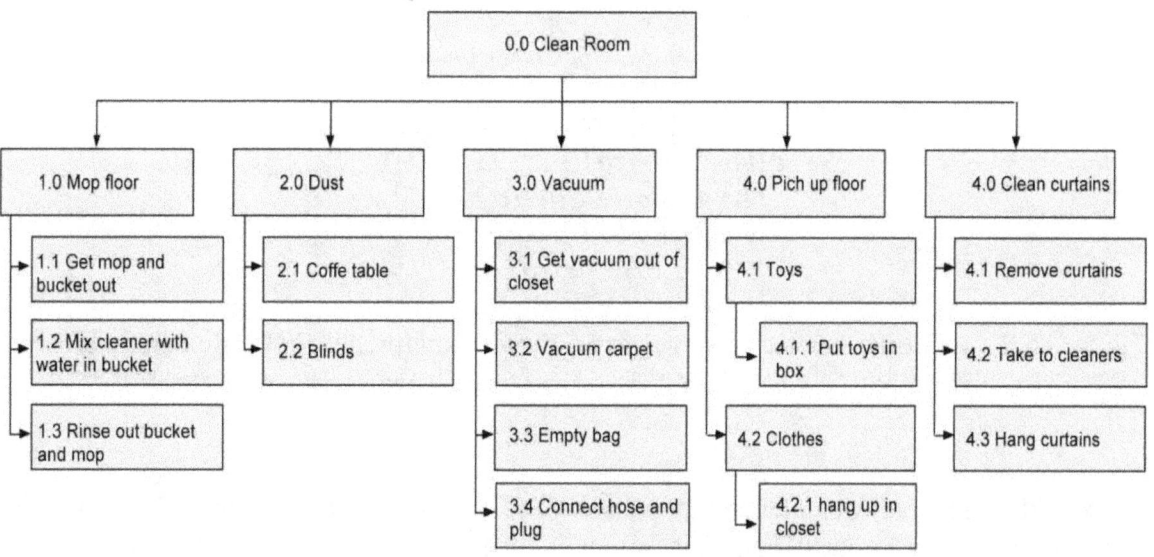

Figure 15.1: A work breakdown structure (WBS) for cleaning a room.
Illustration from Barron & Barron Project Management for Scientists and Engineers, http://cnx.org/content/col11120/1.4/

It is very important to note that when we do a WBS, we do not worry about the sequence in which the work is performed or any dependencies between them. The order of events will be worked out when we develop the schedule. For example, under 3.0 Vacuum (in Figure 11.3), it would be obvious that 3.3 Vacuum carpet would be performed after 3.4 Connect hose and plug. However, you will probably find yourself thinking sequentially, as it seems to be human nature to do so. The main idea of creating a WBS is to capture all of the tasks, irrespective of their order. So if you find yourself and other members of your team thinking sequentially, don't be too concerned, but don't get hung up on trying to diagram the sequence, or you will slow down the process of task identification.

A WBS can be structured any way it makes sense to you and your project. In practice, the chart structure is used quite often (as in the example in Figure 15.1) and it can be composed in an outline form as well (Figure 15.2).

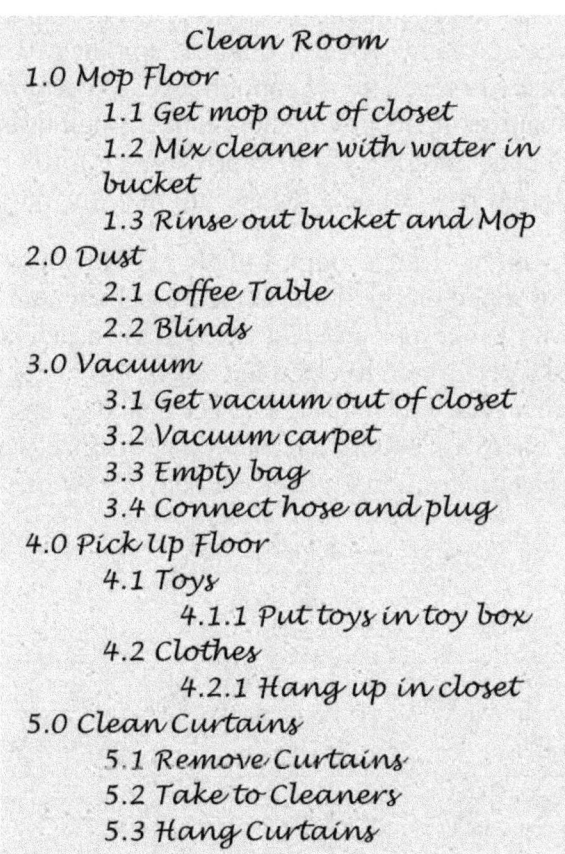

Figure 15.2 An outline format of a work breakdown structure (WBS) for cleaning a room.
Illustration from Barron & Barron Project Management for Scientists and Engineers, http://cnx.org/content/col11120/1.4/

You'll notice that each element at each level of the WBS (either Figure 15.1 or Figure 15.2) is assigned a unique identifier. This unique identifier is typically a number, and it's used to sum and track costs, schedules, and resources associated with WBS elements. These numbers are usually associated with the corporation's chart of accounts, which is used to track costs by category. Collectively, these numeric identifiers are known as ―the code of accounts".

There are also many ways you can organize the WBS. For example, it can be organized by either deliverables or phases. The major deliverables of the project are used as the first level in the WBS. For example, if you are doing a multimedia project the deliverables might include producing a book, CD and a DVD (Figure 15.4).

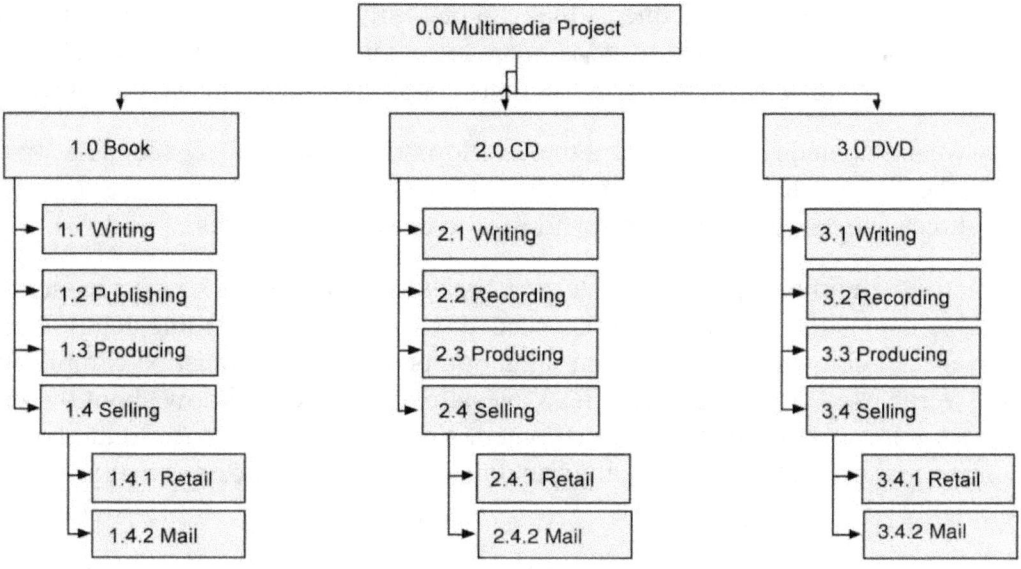

Figure 15.4: An example of a work breakdown structure (WBS) based on project deliverable.
Illustration from Barron & Barron Project Management for Scientists and Engineers, http://cnx.org/content/col11120/1.4/

Many projects are structured or organized by project phases. Each phase would represent the first level of the WBS and their deliverables would be the next level and so on (Figure 15.5).

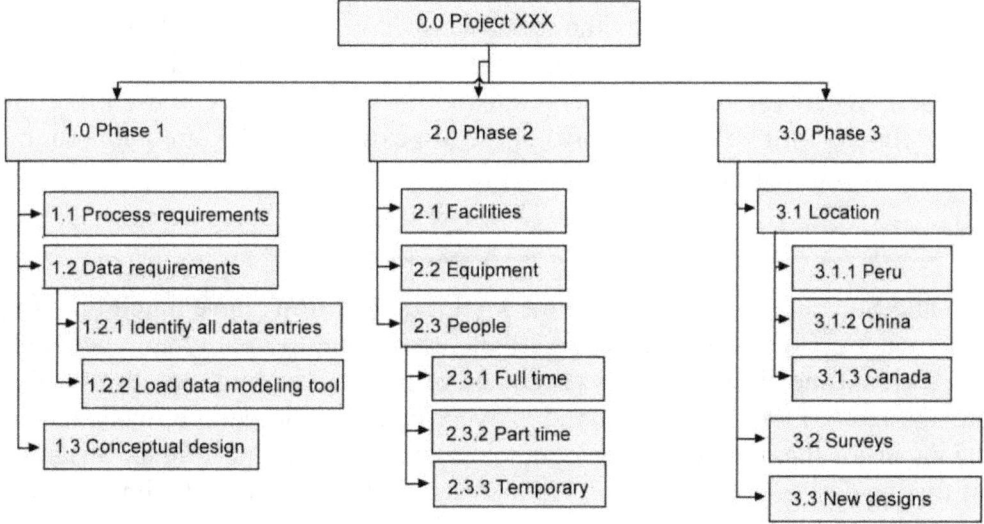

Figure 15.5: An example of a work breakdown structure (WBS) based on project phase.
Illustration from Barron & Barron Project Management for Scientists and Engineers, http://cnx.org/content/col11120/1.4/

As mentioned earlier, the project manager is free to determine the number of levels in the WBS, based on the complexity of the project. You need to include enough levels to accurately estimate project time and costs, but not so many levels that are difficult to distinguish between components. Regardless of the number of levels in a WBS, the lowest level in a WBS is called "a work package".

Work packages are the components that can be easily assigned to one person, or team of people, with clear accountability and responsibility for completing the assignment. The work package level is where time estimates, costs estimates and resource estimates are determined.

Now we are up and running toward the development of our project schedule. In order to develop our schedule, we first need to define the activities, sequence them in the right order, estimate resources and estimate the time it will take to complete the tasks.

The activity definition process is a further breakdown of the work package elements of the WBS. It documents the specific activities needed to fulfill the deliverable detailed in the WBS. These are not deliverables but the individual units of work that must be completed to fulfill the deliverables. Activity definition uses everything we already know about the project to divide the work into activities that can be estimated. You might want to look at all the lessons learned from similar projects your company has done to get a good idea of what you need to do on the current one.

Expert judgment in the form of project team members with prior experience developing project scope statements and WBS can help you define activities. You might also use experts in a particular field to help define tasks, if you were asked to manage a project in a new domain, to help you understand what activities were going to be involved. It could be that you create an activity list and then have the expert review it and suggest changes. Alternatively, you could involve the expert from the very beginning and ask to have an activity definition conversation with him/her before even making your first draft of the list.

Sometimes you start a project without knowing a lot about the work that you'll be doing later. Rolling wave planning lets you plan and schedule only the portion that you know enough about to plan well. When you don't know enough about a project, you can use placeholders for the unknown portions, until you know more. These are extra items that are put at high levels in the WBS to allow you to plan for the unknown.

15.2   A case study

Susan and Steve have decided to tie the knot, but they don't have much time to plan their wedding. They want the big day to be unforgettable, they want to invite many people and provide them a great time. They've always dreamed of a June wedding, but it's already January. Just thinking about all of the details involved is overwhelming. Sometime, when they were choosing the paper for the invitations, the couple realized that they need help. Susan has been dreaming of the big day since she was 12, but it seems that there's so little time for all the tasks to be completed.

	Steve, we need some help.

	Don't worry. My sister's wedding planner was great. Let me give her a call. Steve calls the wedding planner Sally.
	Hello Susan and Steve.
	We want everything to be perfect. My sister said you really saved her wedding. I know she gave you over a year to plan.
	There is so much to do! Invitations, food, guests, and music.
	Oh no, we haven't even booked a place!
	And it has to be done right. We can't print the initiations until we have the menu planned. We can't do the seating arrangements until we have the RSVPs. We aren't sure what kind of band to get for the reception, or should it be a DJ? We're just overwhelmed.
	My sister said you really saved her wedding. I know she gave you over a year to plan.

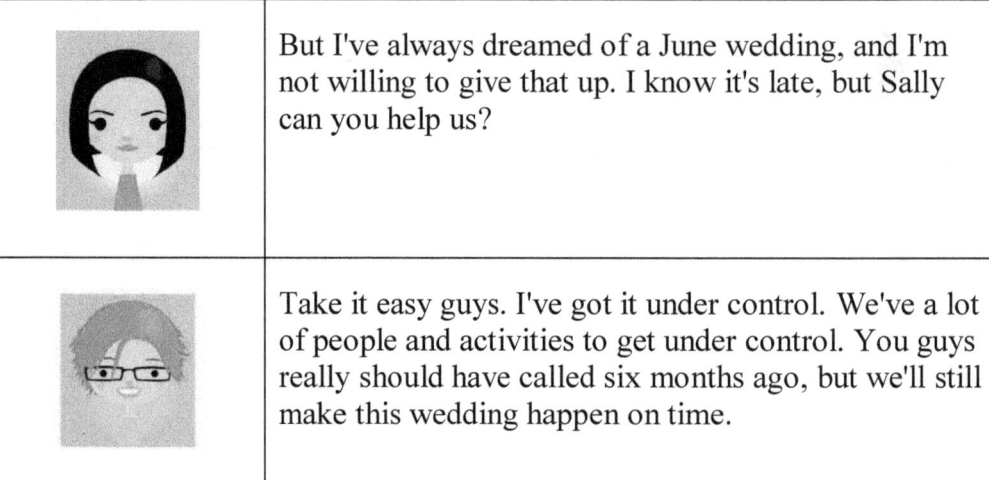

	But I've always dreamed of a June wedding, and I'm not willing to give that up. I know it's late, but Sally can you help us?
	Take it easy guys. I've got it under control. We've a lot of people and activities to get under control. You guys really should have called six months ago, but we'll still make this wedding happen on time.

Illustration from Barron & Barron Project Management for Scientists and Engineers, http://cnx.org/content/col11120/1.4/

Much work has to be done before June. First, Sally figures out what work needs to be done. She starts to put together a to-do-list.

- Invitations

- Flowers

- Wedding Cake

- Dinner Menu

- Band

Since many different people are involved in the making of the wedding, it takes much planning to coordinate all the work in the right order, by the right people at the right time. Initially, Sally was worried that she didn't have enough time to make sure that everything would be done properly. However, she knew that she had some powerful time management tools on her side when she took the job, and these tools would help her to synchronize all the required tasks.

To get started, Sally arranged all the activities in a work breakdown structure. Exercise 15.1 presents part of the WBS Sally made for the wedding.

Exercise 15.1 (Solution in Appendix A)

Arrange the following activities into the WBS to show how the work items decompose into activities.

- Shop for shoes

- Create guest list

- Tailoring and fitting

- Shop for dress

- Find caterer

- Cater the wedding

- Wait for RSVPs

- Mail the invitations

- Finalize the menu

- Print the invitations

- Choose the bouquet

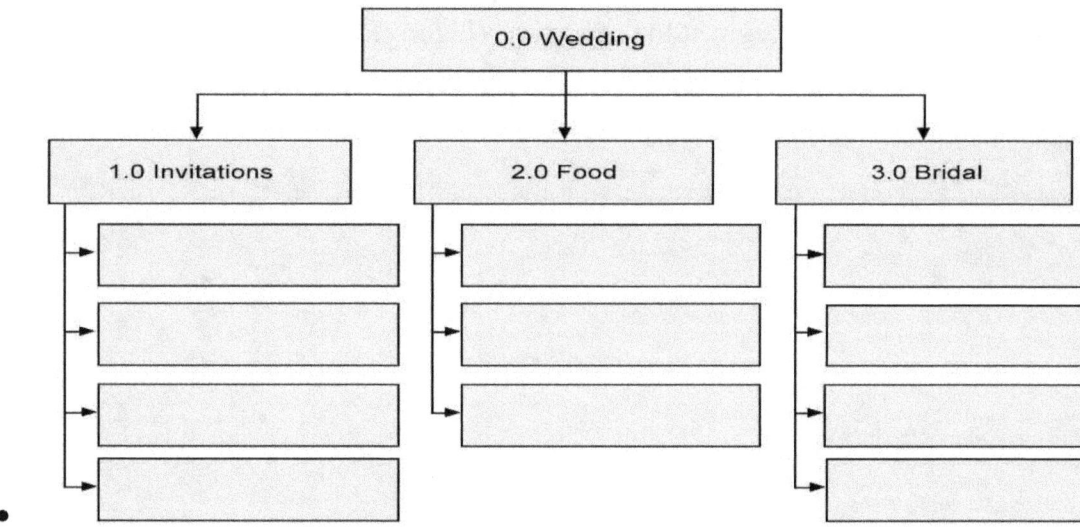

-

Figure 15.4: An example of a work breakdown structure (WBS) based on project phase.
Illustration from Barron & Barron Project Management for Scientists and Engineers, http://cnx.org/content/col11120/1.4/

15.3   Activity Definition

Now that the activity definitions for the work packages have been completed, the next task is to complete the activity list. The project activity list is a list of everything that needs to be done to complete your project, including all the activities that must be accomplished to deliver the work package. Next you want to define the activity attributes. Here's where the description of each activity is kept. All of the information you need to figure out; the order of the work should be here too. So any predecessor activities, successor activities or constraints should be listed in the attributes along with descriptions and any other information about resources or time that you need for planning. The three main kinds of predecessors are finish-to-start (FS), start-to-start (SS) and finish-to-finish (FF). The most common kind of predecessor is the finish-to-start. It means that one task needs to be completed before another one can start. When you think of predecessors, this is what you usually think of; one thing needs to end before the next can begin. It's called finish-to-start because the first activity's finish leads into the second activity's start (Figure 15.5).

Figure 15.5: An example of a finish-to-start (FS) predecessor.
Illustration from Barron & Barron Project Management for Scientists and Engineers, http://cnx.org/content/col11120/1.4/

The start-to-start predecessor is a little less common, but sometimes you need to coordinate activities so they begin at the same time (Figure 15.6).

Figure 15.6: An example of a start-to-start (SS) predecessor.
Illustration from Barron & Barron Project Management for Scientists and Engineers, http://cnx.org/content/col11120/1.4/

In the finish-to-finish predecessor it shows activities that finish at the same time (Figure 15.7).

Figure 15.7: An example of a finish-to-finish (FF) predecessor.
Illustration from Barron & Barron Project Management for Scientists and Engineers, http://cnx.org/content/col11120/1.4/

It is possible to have to have start-to-finish (SF) predecessors. This happens when activities require that another task be started before the successor task can finish. An example might be that the musicians cannot finish playing until the guests have started leaving the ceremony. In addition there are some particular types of predecessors that must be considered.

15.3.1   External Predecessors

Sometimes your project will depend on things outside the work you're doing. For the wedding, we are depending on the wedding party before us to be out of the reception hall in time for us to decorate. The decoration of the reception hall then depends on that as an external predecessor.

15.3.2   Discretionary Predecessors

These are usually process or procedure driven or best practice techniques based on past experience. In the wedding example: Steve and Susan want the bridesmaids to arrive at the reception before the couple arrives. There's no necessity; it is just a matter of preference.

15.3.3   Mandatory Predecessors

You can't address an invitation that hasn't been printed yet. So, printing invitations is a mandatory predecessor for addressing them. Mandatory predecessors are the kinds that have to exist just because of the nature of the work.

15.4   Leads and Lags

Sometimes you need to give some extra time between activities. Lag time is when you purposefully put a delay between the predecessor task and the successor. For example, when the bride and her father dance, the others wait awhile before they join them (Figure 15.8).

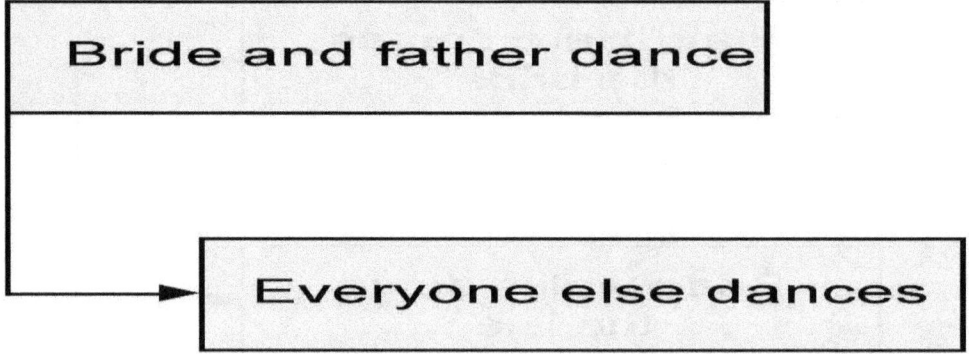

Figure 15.8: A lag means making sure that one task waits a while before it gets started.
Illustration from Barron & Barron Project Management for Scientists and Engineers, http://cnx.org/content/col11120/1.4/

Lead time is when you give a successor task some time to get started before the predecessor finishes (Figure 15.9). So you might want the caterer preparing dessert an hour before everybody is eating dinner.

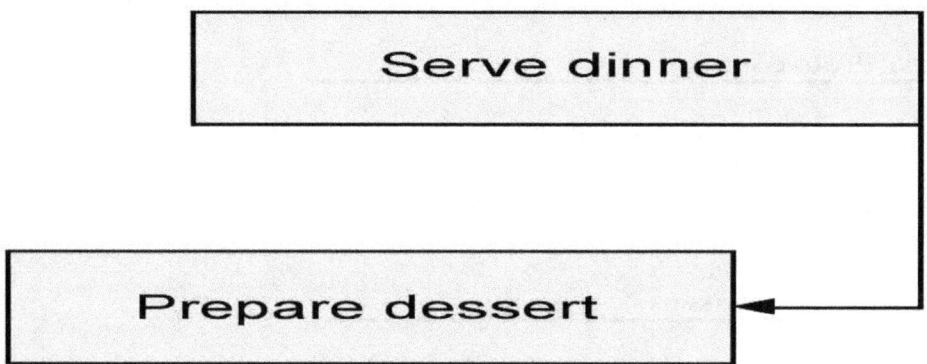

Figure 15.9: A lead is when you let a task get started before its predecessor is done.
Illustration from Barron & Barron Project Management for Scientists and Engineers, http://cnx.org/content/col11120/1.4/

15.4   Milestones

All of the important checkpoints of your project are tracked as milestones. Some of them could be listed in your contract as requirements of successful completion; some could just be significant points in the project that you want to keep track of. The milestone list needs to let everyone know which are required and which are not.

Some milestones for Susan and Steve's wedding might be:

- Invitations sent

- Menu finalized

- Location booked

- Bridesmaids' dresses fitted

As you figure out which activities will need to be done, you may realize that the scope needs to change. When that happens, you need to create a change request and send it through the change control system. So back to our couple and their nuptial plan.

	We just got the programs back from the printer and they're all wrong.
	The quartet cancelled. They had another wedding that day.
	Aunt Jane is supposed to sing at the service, but after what happened at her uncle's funeral, I think I want someone else to do it.
	Should we really have a pan flute player? I'm beginning to think it might be overkill.
	Apparently! Maybe we should hold off on printing the invitations until these things are worked out.
	OK, let's think about exactly how we want to do this. I think we need to be sure about how we want the service to go before we do any more printing.

15.5   The Activity Sequencing Process

Now that we know what we have to do to make the wedding a success, we need to focus on the order of the work. Sally sat down with all of the activities she had defined for the wedding and decided to figure out exactly how they needed to happen. That's where she used the activity sequencing process.

The activity attribute list Sally created had most of the predecessors and successors necessary written in it. This is where she thought of what comes first, second, third, etc. Sally's milestone list had major pieces of work written down and there were a couple of changes to the scope she had discovered along the way that were approved and ready to go.

Example 15.5 Milestone list: Steve and Susan had asked that the invitations be printed at least three months in advance to be sure that everyone had time to RSVP. That's a milestone on Sally's list.

Example 15.6 Change request: When Sally realized that Steve and Susan were going to need another limo to take the bridesmaids to the reception hall, she put that change through change control-including running everything by Susan's mother - and it was approved.

15.6   Creating the Network Diagram

The first step in developing the schedule is to develop a network diagram of the WBS work packages. The network diagram is a way to visualize the interrelationships of project activities. Network diagrams provide a graphical view of the tasks and how they relate to one another. The tasks in the network are the work packages of the WBS. All of the WBS tasks must be included in the network because they have to be accounted for in the schedule. Leaving even one task out of the network could change the overall schedule duration, estimated costs, and resource allocation commitments.

The first step is to arrange the tasks from your WBS into a sequence (Figure 15.12). Some tasks can be accomplished at any time throughout the project where other tasks depend on input from another task or are constrained by time or resources.

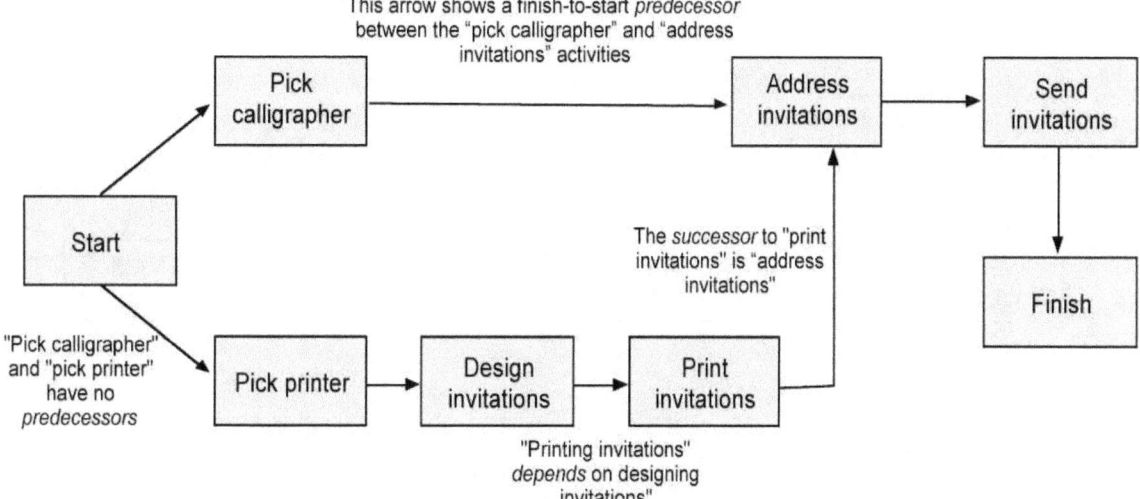

Figure 15.12: The relationship between the work breakdown structure (WBS) and the network diagram.

Illustration from Barron & Barron Project Management for Scientists and Engineers, http://cnx.org/content/col11120/1.4/

The WBS is not a schedule, but it is the basis for it; the network diagram is a schedule but is used primarily to identify key scheduling information that ultimately goes into user friendly schedule formats, such as milestone and Gantt charts.

The network diagram provides important information to the project team. It provides information about how the tasks are related (Figure 15.12), where the risk points are in the schedule, how long it will take as currently planned to finish the project, and when each task needs to begin and end.

In our wedding planner example, Sally would look for relationships between tasks and determine what can be done in parallel and what activities need to wait for others to complete. As an example, Figure 15.13 shows how the activities involved in producing the invitations depend on one another. Showing the activities in rectangles and their relationships as arrows is called a precedence diagramming method (PDM). This kind of diagram is also called an activity-on-node (AON) diagram.

Another way to show how tasks relate is with the activity-on-arrow (AOA). Although activity-on-node (AON) is more commonly used and is supported by all project management programs, PERT is the best-known AOA-type diagram and is the historical basis of all network diagramming. The main difference is the AOA diagram is traditionally drawn using circles as the nodes, with nodes representing the beginning and ending points of the arrows or tasks. In the AOA network, the arrows represent the activities or tasks (Figure 15.14).

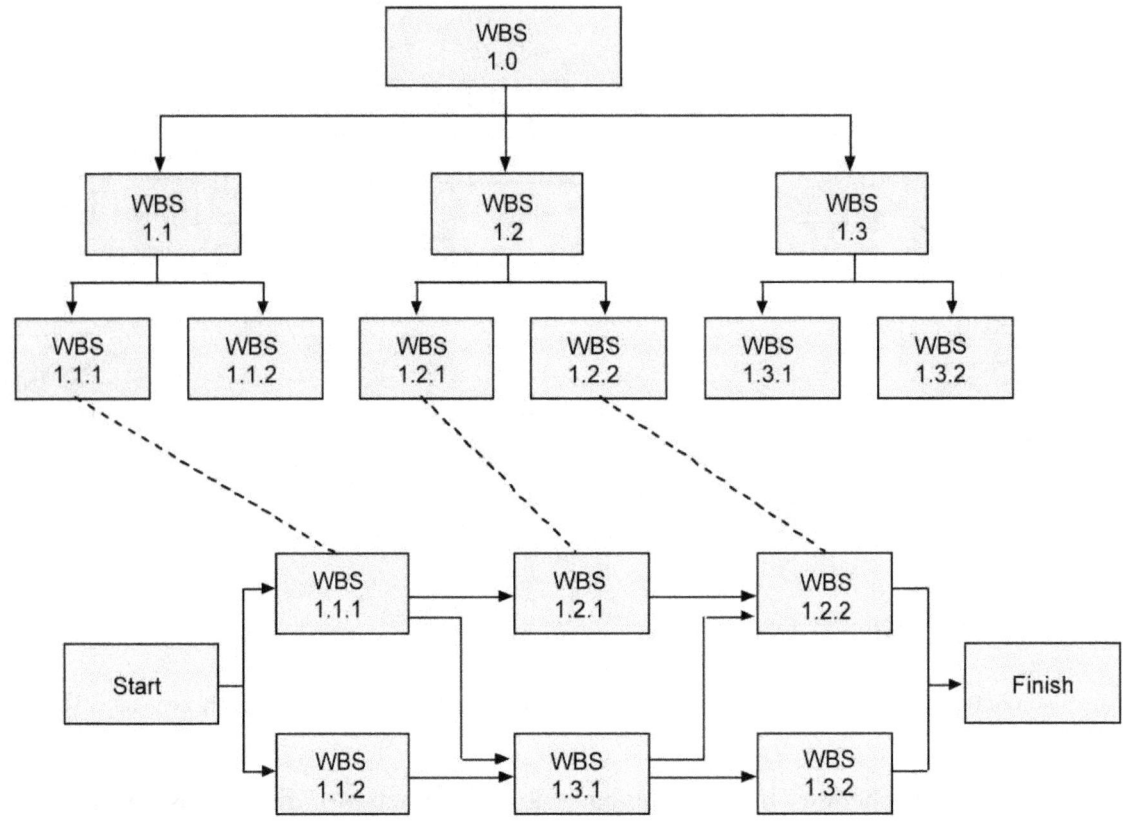

Figure 15.13: An example of an activity on node (AON) diagram.
Illustration from Barron & Barron Project Management for Scientists and Engineers, http://cnx.org/content/col11120/1.4/

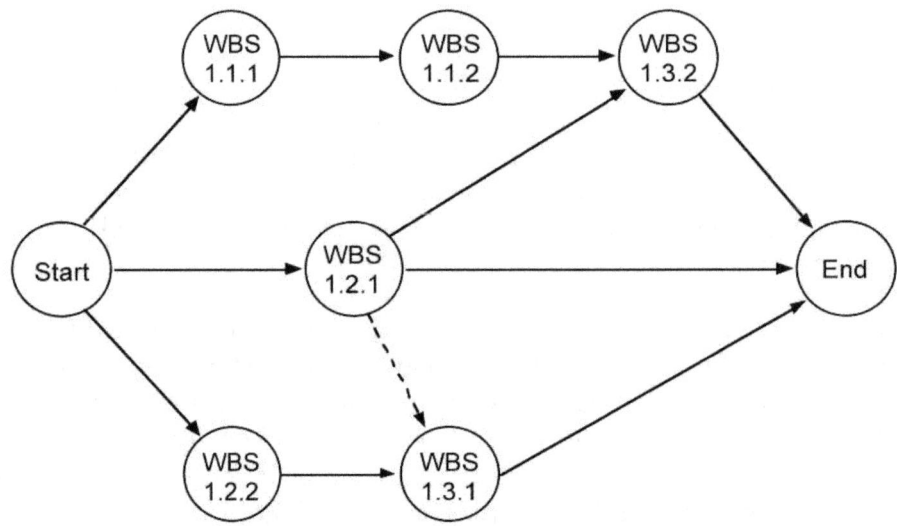

Figure 15.14.: An example of an activity narrow (AOA) network diagram.
Illustration from Barron & Barron Project Management for Scientists and Engineers, http://cnx.org/content/col11120/1.4/

All network diagrams have the advantages as showing task interdependencies, start and end times, and the critical path (the longest path through the network) but the AOA network also has some disadvantages that limit the use of the method.

The three major disadvantages of the AOA method are:

- The AOA network can only show finish to start relationships. It is not possible to show lead and lag except by adding or subtracting time, which makes project tracking difficult.

- There are instances when dummy activities can occur in an AOA network. Dummy activities are activities that show the dependency of one task on other tasks but for other than technical reasons. For example, a task may be dependent on another because it would be more cost effective to use the same resources for the two; otherwise the two tasks could be accomplished in parallel. Dummy activities do not have durations associated with them. They simply show that a task has some kind of dependence on another task.

- AOA diagrams are not as widely used as AON simply because the latter are somewhat simpler to use and all project management software programs can accommodate AON networks, whereas not all can accommodate AOA networks.

In our case study it is clear that Steve and Susan have resource problems. Getting a handle on all of the tasks that have to be done is a great start, but it's not enough to know the tasks and the order they come in. Before you can put the final schedule together, you need to know who is going to do each job, and the things they need available to them in order to do it.

—We've got so much to do! Invitations, catering, music... and I've got no idea who's going to do it all. I'm totally overwhelmed." From this statement it is clear that Susan is worried about human resources. In comparison, Steve realizes that not all resources are people.

And it's not just people. We need food, flowers, a cake, a sound system, and a venue. How do we get a handle on this?

Resources are people, equipment, place, money, or anything else that you need in order to do all of the activities that you planned for. Every activity in your activity list needs to have resources assigned to it. Before you can assign resources to your project, you need to know their availability. Resource availability includes information about what resources you can use on your project, when they're available to you, and conditions of their availability. Don't forget that some resources like consultants or training rooms have to be scheduled in advance, and they might only be available at certain times. You'll need to know this before you can finish planning your project. If you are starting to plan in January, a June wedding is harder to plan than one in December, because the wedding halls are all booked up in advance. That is clearly a resource constraint. You'll also need the activity list that you created earlier, and you'll need to know about how your organization typically handles resources. Once you've got a handle on these things, you're set for resource estimation.

16.1   Estimating the Resources

The goal of activity resource estimating is to assign resources to each activity in the activity list. There are five tools and techniques for estimating activity resources.

Expert judgment means bringing in experts who have done this sort of work before and getting their opinions on what resources are needed (Figure 11.15).

Alternative analysis means considering several different options for how you assign resources. This includes varying the number of resources as well as the kind of resources you use. Many times, there's more than one way to accomplish an activity and alternative analysis helps decide among the possibilities.

Published estimating data is something that project managers in a lot of industries use to help them figure out how many resources they need. They rely on articles, books, journals, and periodicals that collect, analyze, and publish data from other people's projects.

Project management software such as Microsoft project will often have features designed to help project managers estimate resource needs and constraints and find the best combination of assignments for the project.

Bottom-up estimating means breaking down complex activities into pieces and working out the resource assignments for each piece. It is a process of estimating individual activity resource need or cost and then adding these up together to come up with a total estimate. Bottom-up estimating is a very accurate means of estimating, provided the estimates at the schedule activity level are accurate. However, it takes a considerable amount of time to perform bottom-up estimating because every activity must be accessed and estimated accurately to be included in the bottom-up calculation. The smaller and more detailed the activity, the greater the accuracy and cost of this technique.

In each of the following scenarios of planning Steve and Susan's wedding, determine which of the five activity resource estimation tools and techniques is being used.

Solutions to these exercises are in Appendix A.

Exercise 16.1 Sally has to figure out what to do for the music at Steve and Susan's wedding. She considers using a DJ, a rock band, or a string quartet.
Exercise 16.2 The latest issue of Wedding Planner's Journal has an article on working with caterers. It includes a table that shows how many waiters work with varied guest-list sizes.
Exercise 16.3 There's a national wedding consultant who specializes in Caribbean themed weddings. Sally gets in touch with her to ask about menu options.
Exercise 16.4 Sally downloads and fills out a specialized spreadsheet that a project manager developed to help with wedding planning.
Exercise 16.5 There's so much work that has to be done to set up the reception hall that Sally has to break it down into five different activities in order to assign jobs.
Exercise 16.6 Sally asks Steve and Susan to visit several different caterers and sample various potential items for the menu.
Exercise 16.7 Sally calls up her friend who knows specifics of the various venues in their area for advice on which one would work best.

16.2   Estimating Activity Durations

Once you're done with activity resource estimating, you've got everything you need to figure out how long each activity will take. That's done in a process called activity duration estimating. This is where you look at each activity in the activity list, consider its scope and resources, and estimate how long it will take to perform.

Estimating the duration of an activity means starting with the information you have about that activity and the resources that are assigned to it, and then working with the project team to come up with an estimate. Most of the time you'll start with a rough estimate and then refine it to make it more accurate. You'll use these five tools and techniques to create the most accurate estimates:

- Expert judgment will come from your project team members who are familiar with the work that has to be done. If you don't get their opinion, then there's a huge risk that your estimates will be wrong.

- Analogous estimating is when you look at similar activities from previous projects t and look at how long they took before. But this only works if the activities and resources are similar.

- Parametric estimating means plugging data about your project into a formula, spreadsheet, database, or computer program that comes up with an estimate. The software or formula that you use for parametric estimating is based on a database of actual durations from past projects.

- Three-point estimates are when you come up with three numbers: a realistic estimate that's most likely to occur, an optimistic one that represents the best-case scenario, and a pessimistic one that represents the worst-case scenario. The final estimate is the weighted average of the three.

- Reserve analysis means adding extra time to the schedule (called a contingency reserve or a buffer) to account for extra risk.

In each of the following scenarios for planning Steve and Susan's wedding, determine which of the five activity duration estimation tools and techniques is being used. Solutions are in Appendix A.

Exercise 16.8 There are two different catering companies at the wedding. Sally asks the head chef at each of them to give her an estimate of how long it will take each of them to do the job.

Exercise 16.9 There's a spreadsheet Sally always uses to figure out how long it takes guest to RSVP. She enters the number of guests and their zip codes, and it calculates estimates for her.

Exercise 16.10 Sally's done four weddings that are very similar to Steve and Susan's, and in all four of them it took exactly the same amount of time for the caterers to set up the reception hall.

The activity duration estimates are an estimate of how long each activity in the activity list will take. This is a quantitative measure usually expressed in hours, weeks, days, or months. Any work period is fine, and you'll use different work periods for different jobs. A small job (like booking a DJ) may just take a few hours; a bigger job (like catering-including deciding on a menu, ordering ingredients, cook food and serving guests on the big day) could take days.

Another thing to keep in mind when estimating the duration of the activities, is determining the effort involved. Duration is the amount of the time that an activity takes, while effort if the total number of person-hours that are expended. If it takes two people six hours to carve the ice sculpture for the centerpiece of a wedding, the duration is six hours. But if two people worked on it for the whole time, it took twelve person-hours of effort to create.

You'll also learn more about the specific activities while you're estimating them. That's something that always happens. You have to really think through all of the aspects of a task in order to estimate it. As you learn more about the specific activities remember to update the activity attributes.

If we go back to our case study of the wedding, we can see that while Sally has got a handle on how long things are going to take, she still has some work to do before she's got the whole project under control. Steve and Susan know where they want to get married, and they've got the place booked now. But, what about the caterer? They have no idea who's going to be providing food. And what about the band they want? Will the timing with their schedule work out?

If the caterers come too early, the food will sit around under heat lamps. But, if they come too late, then the band won't have time to play. I just don't see how we'll ever work this out.

It's not easy to plan for a lot of resources when they have tight time restrictions and overlapping constraints. How do you figure out a schedule that makes everything fit together? You're never going to have the complete resource picture until your done building the schedule. And the same goes for your activity list and duration estimates too! It's only when you lay out the schedule that you'll figure out that some of your activities and durations didn't quite work.

16.3   Project Schedule

The project schedule should be approved and signed off by stakeholders and functional managers. This ensures they have read the schedule, understand the dates and resource commitments, and will cooperate. You'll also need to obtain confirmation that resources will be available as outlined in the schedule. The schedule cannot be finalized until you receive approval and commitment for the resource assignments outlined in it.

Once the schedule is approved, it will become your baseline for the remainder of the project. Project progress and task completion will be monitored and tracked against the project schedule to determine if the project is on course as planned.

The schedule can be displayed in a variety of ways, some of which are variations of what you have already seen. Project schedule network diagrams will work as schedule diagrams when you add the start and finish dates to each activity. These diagrams usually show the activity dependencies and critical path.

The critical path method is an important tool for keeping your projects on track. Every network diagram has something that is called the critical path. It's the string of activities that, if you add up all of the durations, is longer than any other path through the network. It usually starts with the first activity in the network and usually ends with the last one.

	Aunt Jane is a vegetarian. That won't be a problem, right?
	Well, let's see. What menu did we give the caterers?
	We didn't give it to them yet; because we won't have the final menu until everyone RSVPs and lets we know which entrée they want.
	But they can't RSVP because we haven't sent out the invitations! What's holding that up?
	We're still waiting to get them back from the printer. We can't send them out if we don't have them yet!
	Oh no! I still have to tell the printer what to print on the invitations, and what paper to use.

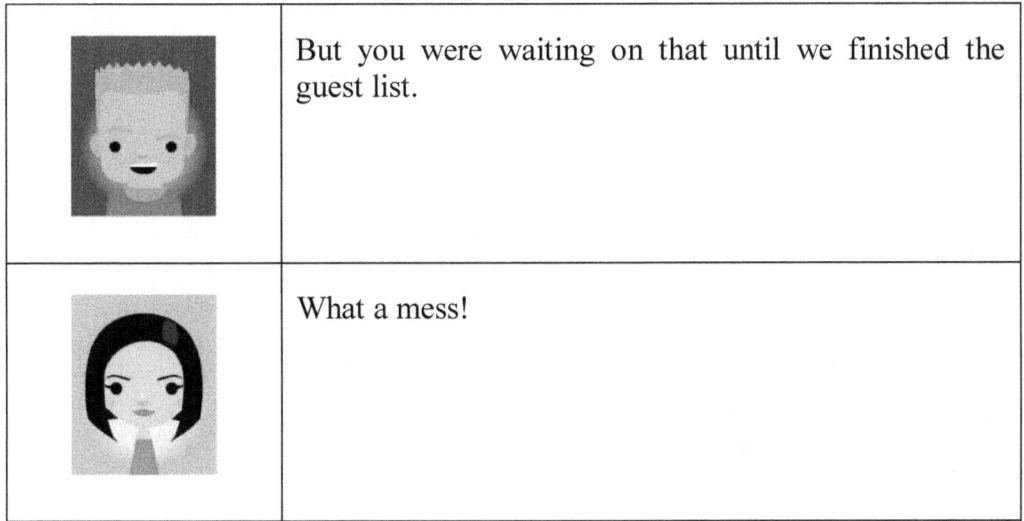	But you were waiting on that until we finished the guest list.
	What a mess!

Steve thought Aunt Jane being a vegetarian was just a little problem. But it turns out to be a lot bigger than either Steve or Susan realized at first. How'd a question about one guest's meal lead to such a huge mess?

The reason that the critical path is critical is that every single activity on the path must finish on time in order for the project to come in on time. A delay in any one of the critical path activities will cause the entire project to be delayed (Figure 16.1).

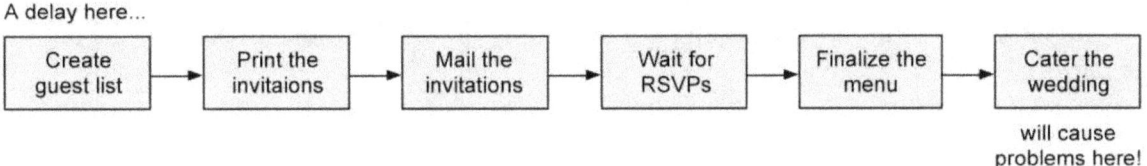

Figure 16.1: An example of problems that can be caused within the critical path.

Knowing where your critical path is can give you a lot of freedom. If you know an activity is not on the critical path, then you know a delay in that activity may not necessarily delay the project. This can really help you handle emergency situations. Even better, it means that if you need to bring your project in earlier than was originally planned, you know that by adding resources to the critical path will be much more effective than adding them elsewhere.

It's easy to find the critical path in any project. Of course, on a large project with dozens or hundreds of tasks, you'll probably use software like Microsoft Project to find the critical path for you. But when it does, it's following the same exact steps that are followed here.

Step 1. Start with a network diagram.

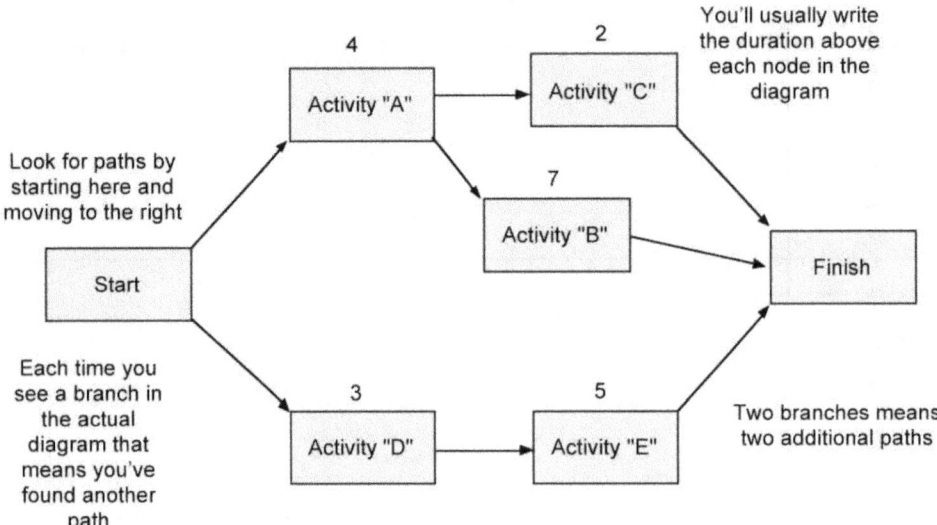

Illustration from Barron & Barron Project Management for Scientists and Engineers,
http://cnx.org/content/col11120/1.4/

Step 2. Find all the paths in the diagram. A path is any string of activities that goes from the start of the project to the end.

Start → Activity "A" → Activity "B" → Finish

Start → Activity "A" → Activity "C" → Finish

Start → Activity "D" → Activity "E" → Finish

Step 3. Find the duration of each path by adding up the durations of each of the activities on the path.

Start → Activity "A" → Activity "B" → Finish = 4 + 7 = 11

Start → Activity "A" → Activity "C" → Finish = 4 + 2 = 6

Start → Activity "D" → Activity "E" → Finish = 3 + 5 = 8

Step 4. The first path has a duration of 11, which is longer than the other paths, so it's the critical path.

The schedule can also be displayed using a Gantt chart (Figure 16.2). Gantt charts are easy to read and commonly used to display schedule activities. Depending on the software you use to display the Gantt chart, it might also show activity sequences, activity start and end dates, resource assignments, activity dependencies, and the critical path. Gantt charts are also known as bar charts.

	Task Name	Duration	Start	Finish
1	Design Server database structure	2 days	Sat 4/7/01	Sun 4/8/01
2	Implement Database Module	6 days	Mon 4/9/01	Sat 4/14/01
3	Implement Database Utility Module	4 days	Sun 4/15/01	Wed 4/18/01
4	Implement SuperUserManager Module	2 days	Thu 4/19/01	Fri 4/20/01
5	Implement I/O modules	7 days	Sun 4/8/01	Sat 4/14/01
6	Research RMILite and Ninja	4 days	Sun 4/15/01	Wed 4/18/01
7	Design Palm database structure	10 days	Thu 4/19/01	Sat 4/28/01
8	Implement Meeting Module	2 days	Mon 4/9/01	Tue 4/10/01
9	Implement Meeting Manager Module	2 days	Wed 4/11/01	Thu 4/12/01
10	Implement Open Meeting Monitor module	3 days	Fri 4/13/01	Sun 4/15/01
11	Implement Schedule Module	3 days	Wed 4/18/01	Fri 4/20/01
12	Implement Composite Schedule Module	4 days	Sat 4/21/01	Tue 4/24/01
13	Implement Authentication Manager Module	1 day	Thu 4/26/01	Thu 4/26/01
14	Implement Log Manager Module	1 day	Fri 4/27/01	Fri 4/27/01
15	Research OSKI	1 day	Sat 4/7/01	Sat 4/7/01
16	Implement User Synchronization Module	9 days	Sun 4/8/01	Mon 4/16/01
17	Implement Server Synchronization Module	9 days	Wed 4/18/01	Thu 4/26/01
18	Implement PalmOS GUI screen modules	9 days	Sat 4/7/01	Sun 4/15/01
19	Implement Client Rendezvous Application Module	4 days	Thu 4/19/01	Sun 4/22/01
20	Implement UI Manager Module	3 days	Mon 4/23/01	Wed 4/25/01
21	Design Intergration Tests	7 days	Tue 4/24/01	Mon 4/30/01
22	Intergration Testing	7 days	Tue 5/1/01	Mon 5/7/01
23	Blackbox Testing	3 days	Tue 5/8/01	Thu 5/10/01
24	User Testing	1 day	Fri 5/11/01	Fri 5/11/01
25	User Manual	1 day	Sat 5/12/01	Sat 5/12/01
26	Presentation Preperation	1 day	Mon 5/14/01	Mon 5/14/01

Figure 16.2: An example of a Gantt chart.
Illustration from Barron & Barron Project Management for Scientists and Engineers,
http://cnx.org/content/col11120/1.4/

Chapter 17: Budget Planning

Every project boils down to money. If you had a bigger budget, you could probably get more people to do your project more quickly and deliver more. That's why no project plan is complete until you come up with a budget (Figure 11.18). But no matter whether your project is big or small, and no matter how many resources and activities are in it, the process for figuring out the bottom line is always the same.

It is important to come up with detailed estimates of all the project costs. Once this is obtained, add up the cost estimates into a budget plan. It is now possible to track the project according to that budget while the work is ongoing.

A lot of times you come into a project and there is already an expectation of how much it will cost or how much time it will take. When you make an estimate really early in the project and you don't know much about it, that estimate is called a rough order of magnitude estimate (or a ballpark estimate). It's expected that this estimate will become more refined as time goes on and you learn more about the project. Here are some more tools and techniques used to estimate cost:

Determine resource cost rates: People who will be working on the project all work at a specific rate. Any materials you will use to build the project (like wood or wiring) will be charged at a rate too. This just means figuring out what the rate for labor and materials will be.

Vendor bid analysis: Sometimes you will need to work with an external contractor to get your project done. You might even have more than one contractor bid on the job. This tool is all about evaluating those bids and choosing the one you will go with.

Reserve analysis: You need to set aside some money for cost overruns. If you know that your project has a risk of something expensive happening, better to have some cash lying around to deal with it. Reserve analysis means putting some cash away just in case.

Cost of quality: You will need to figure the cost of all your quality related activities into the overall budget, too. Since it's cheaper to find bugs earlier in the project than later, there are always quality costs associated with everything your project produces. Cost of quality is just a way of tracking the cost of those activities and is how much money it takes to do the project right.

Once you apply all the tools in this process, you will arrive at an estimate for how much your project will cost. It's always important to keep all of your supporting estimate information, too. That way, you know the assumptions you made when you were coming up with your numbers. Now you are ready to build your budget plan.

Procurement management follows a logical order. First, you plan what you need to contract; then you plan how you'll do it. Next, you send out your contract requirements to sellers. They bid for the chance to work with you. You pick the best one, and then you sign the contract with them. Once the work begins, you monitor it to make sure that the contract is being followed. When the work is done, you close out the contract and fill out all the paperwork.

You will need to start with a plan for the whole project. You need to think about all of the work that you will contract out for your project before you do anything else. You will want to plan for any purchases and acquisitions. Here's where you take a close look at your needs, to be sure that you really need to create a contract. You figure out what kinds of contracts make sense for your project, and you try to define all of the parts of your project that will be contracted out.

Contract planning is where you plan out each individual contract for the project work. You work out how you to manage the contract, what metrics it will need to meet to be considered successful, how you'll pick a seller, and how you'll administer the contract once the work is happening.

The procurement management plan details how the procurement process will be managed. It includes the following information:

- The types of contracts you plan to use, and any metrics that will be used to measure the contractor's performance.

- The planned delivery dates for the work or products you are contracting.

- The company's standard documents you will use.

- How many vendors or contractors are involved and how they will be managed.

- How purchasing may impact the constraints and assumptions of the project plan.

- Coordination of purchasing lead times with the development of the project schedule.

- Identification of prequalified sellers (if known).

The procurement management plan like all other management plans becomes a subsidiary of the project management plan. Some tools and techniques you may use during the procurement planning stage include make or buy analysis and defining the contract type.

17.1   Make or Buy Analysis

This means figuring out whether or not you should be contracting the work or doing it yourself. It could also mean deciding whether to build a solution to your problem or buy one that is already available. Most of the same factors that help you make every other major project decision will help you with this one. How much does it cost to build it as opposed to buy it? How will this decision affect the scope of your project? How about project schedule? Do you have time to do the work and still meet your commitments? As you plan out what you will and won't contract, you need to have thought through your reasoning pretty carefully.

There are some resources (like heavy equipment) that your company can buy, rent, or lease depending on the situation. You'll need to examine leasing versus buying costs and determine the best way to go forward.

17.2   Contract Types

You should know a little bit about the major kinds of contracts available to you so that you choose the one that creates the most fair and workable deal for you and the contractor. Some contracts are fixed price: no matter how much time or effort goes into them, you always pay the same (Figure 17.1). Some are cost reimbursable also called cost plus (Figure 17.2). This is where the seller charges you for the cost of doing the work plus some fee or rate. The third major kind of contract is time and materials (Figure 17.3). That's where the buyer pays a rate for the time spent working on the project and also pays for all the materials used to do the work.

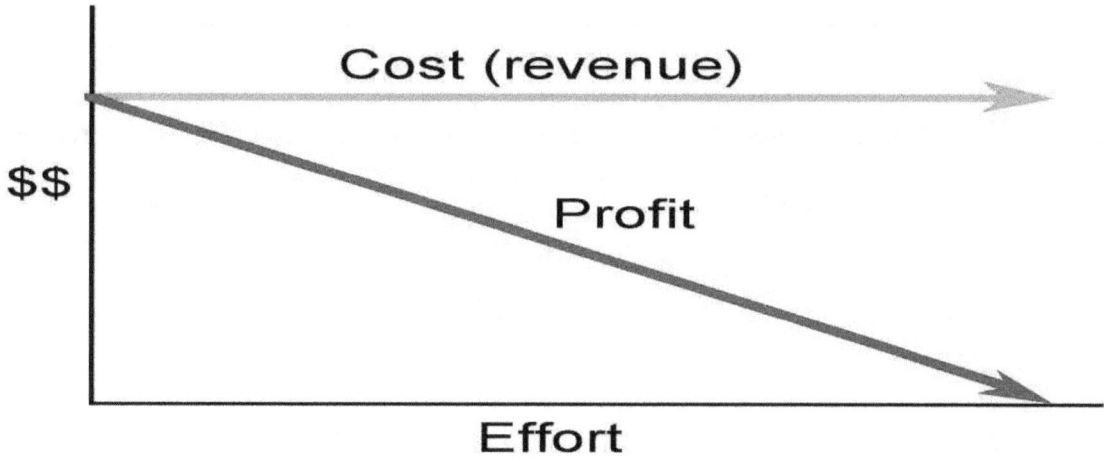

Figure 17.1: A fixed price contract the cost (or revenue to the vendor) is constant regardless of effort applied or delivery date.

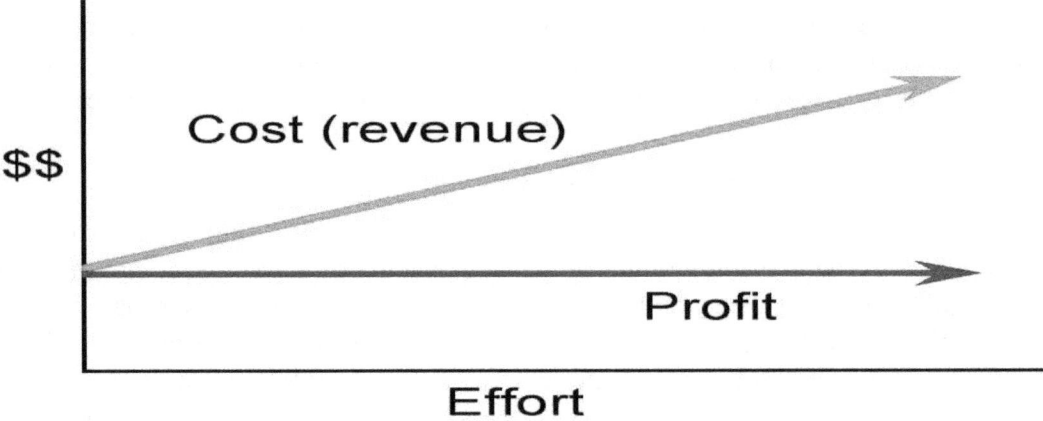

Figure 17.2: In a cost reimbursable or cost plus contract, the seller is guaranteed a specific fee.

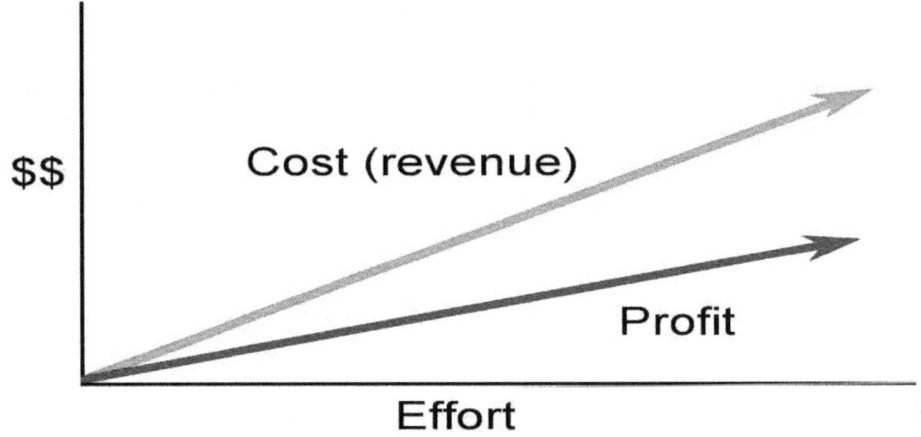

Figure 17.3: In a time and materials contract the cost (or revenue to the vendor) increases with increased effort.

Even the most carefully planned project can run into trouble. No matter how well you plan, your project can always run into unexpected problems. Team members get sick or quit, resources that you were depending on turn out to be unavailable, even the weather can throw you for a loop (For example, Hurricane Ike). So does that mean that you're helpless against unknown problems? No! You can use risk planning to identify potential problems that could cause trouble for your project, analyze how likely they'll occur, take action to prevent the risks you can avoid, and minimize the ones that you can't.

A risk is any uncertain event or condition that might affect your project. Not all risks are negative. Some events (like finding an easier way to do an activity) or conditions (like lower prices for certain materials) can help your project. When this happens, we call it an opportunity; but it's still handled just like a risk.

There are no guarantees on any project. Even the simplest activity can turn into unexpected problems. Any time there's anything that might occur on your project and change the outcome of a project activity, we call that a risk. A risk can be an event (like a hurricane) or it can be a condition (like an important part being unavailable). Either way, it's something that may or may not happen ...but if it does, then it will force you to change the way you and your team will work on the project.

If your project requires that you stand on the edge of a cliff, then there's a risk that you could fall (Figure 18.1). If it's very windy out or if the ground is slippery and uneven, then falling is more likely.

| Your project | Avoid | Mitigate | Transfer | Accept |

Figure 18.1 Potential ways to handle risk in a project.

Illustration from Barron & Barron Project Management for Scientists and Engineers, http://cnx.org/content/col11120/1.4/

When you're planning your project, risks are still uncertain: they haven't happened yet. But eventually, some of the risks that you plan do happen. And that's when you have to deal with them. There are four basic ways to handle a risk.

1. Avoid: The best thing that you can do with a risk is to avoid it. If you can prevent it from happening, it definitely won't hurt your project. The easiest way to avoid this risk is to walk away from the cliff (Figure 18.1), but that may not be an option on this project.

2. Mitigate: If you can't avoid the risk, you can mitigate it. This means taking some sort of action that will cause it to do as little damage to your project as possible (Figure 18.1).

3. Transfer: One effective way to deal with a risk is to pay someone else to accept it for you (Figure 18.1).The most common way to do this is to buy insurance.

4. Accept: When you can't avoid, mitigate, or transfer a risk, then you have to accept it (Figure 18.1). But even when you accept a risk, at least you've looked at the alternatives and you know what will happen of it occurs. If you can't avoid the risk, and there's nothing you can do to reduce its impact, then accepting it is your only choice.

By the time a risk actually occurs on your project, it's too late to do anything about it. That's why you need to plan for risks from the beginning and keep coming back to do more planning throughout the project.

The risk management plan tells you how you're going to handle risk in your project. It documents how you'll access risk on the project, who is responsible for doing it, and how often you'll do risk planning (since you'll have to meet about risk planning with your team throughout the project.)

The plan has parts that are really useful for managing risks.

Here are some risk categories that you'll use to classify your risks. Some risks are technical, like a component that might turn out to be difficult to use. Others are external, like changes in the market or even problems with the weather.

Risk breakdown structure (RBS) is a great tool for managing your risk categories. It looks like a WBS, except instead of tasks it shows how the risks break down into categories.

It's important to come up with guidelines to help you figure out how big a risk's impact is. The impact tells you how much damage the risk will cause to your project. A lot of projects classify impact on a scale from minimal to severe, or from very low to very high. The plan should also give you a scale to help figure out the probability of the risk. Some risks are very likely; others aren't.

It's not enough to make sure you get it done on time and under budget. You need to be sure you make the right product to suit your stakeholders' needs. Quality means making sure that you build what you said you would and that you do it as efficiently as you can. That means trying not to make too many mistakes and always keeping your project working toward the goal of creating the right product.

Everybody "knows" what quality is. But the way the word is used in everyday life is a little different that how it is used in project management. Just like the triple constraint, scope, cost, and schedule-you manage quality on your project by setting goals and taking measurements. That's why you need to understand the quality levels your stakeholders believe are acceptable, and that your projects meet those targets; just like it needs to meet their budget and schedule goals.

Customer satisfaction is about making sure that the people who are paying for the end product are happy with what they get. When the team gathers requirements for the specification, they try to write down all of the things that the customers want in the product so that you know how to make them happy. Some requirements can be left unstated, too. Those are the ones that are implied by the customer's explicit needs. For example: some requirements are just common sense, like a product that people hold can't be made from toxic chemicals that may kill you. It might not be stated, but it's definitely a requirement.

Fitness to use is about making sure that the product you build has the best design possible to fit the customer's needs. Which would you choose: a product that's beautifully designed, well constructed, solidly built and all around pleasant to look at but does not do what you need, or a product that does what you want despite being really ugly and hard to use? You'll always choose the product that fits your needs, even if it's seriously limited. That's why it's important that the product both does what it is supposed to do and does it well. For example: you could pound in a nail with a screwdriver, but a hammer is better fit for the job.

Conformance to requirements is the core of both customer satisfaction and fitness to use, and is a measure of how well your product does what you intend. Above all, your product needs to do what you wrote down in your requirements document. Your requirements should take into account what will satisfy your customer and the best design possible for the job. That means conforming to both stated and implied requirements.

In the end, your product's quality is judged by whether you built what you said you would build

Quality planning focuses on taking all of the information available to you at the beginning of your project and figuring out how you will measure your quality and prevent defects. Your company should have a quality policy that tells how it measures quality across the organization. You should make sure your project follows the company policy and any governmental rules or regulations on how you need to plan quality for your project.

You need to plan out which activities you're going to use to measure the quality of the product of your project. And you need to be sure the activities you plan are going to pay off in the end. So you'll need to think about the cost of all the quality-related activities you want to do. Then you'll need to set some guidelines for what you going to measure against. Finally, you'll need to design the tests you're going to run when the product is ready to be tested.

19.1   Quality planning tools

The following represents the quality planning tools available to the project manager.

- Cost benefit analysis is looking at how much your quality activities will cost versus how much you will gain from doing them. The costs are easy to measure; the effort and resources it takes to do them are just like any other task on your schedule. Since quality activities don't actually produce a product, it is harder for people to measure the benefit sometimes. The main benefits are less re-work, higher productivity and efficiency and more satisfaction from both the team and the customer.

- Benchmarking means using the results of quality planning on other projects to set goals for your own. You might find that the last project in your company had 20% fewer defects than the one before it. You should want to learn from a project like that and put in practice any of the ideas they used to make such a great improvement. Benchmarks can give you some reference points for judging your own project before you even get started with the work.

- Design of experiments is the list of all the kinds of tests you are going to run on your product. It might list all the kinds of test procedures you'll do, the approaches you'll take, and even the tests themselves. (In the software world, this is called test planning).

- Cost of quality is what you get when you add up the cost of all the prevention and inspection activities you are going to do on your project. It doesn't just include the testing. It includes any time spent writing standards, reviewing documents, meeting to analyze the root causes of defects, re-work to fix the defects once they're found by the team; absolutely everything you do to ensure quality on the project.

- Cost of quality can be a good number to check whether your project is doing well or having trouble. Say your company tracks cost of quality on all of its projects. Then you could tell if you were spending more or less than they are to get your project up to quality standards.

Once you have your quality plan, you know your guidelines for managing quality on your project. Your strategies for monitoring your project quality should be included in the plan, as well as the reasons for all the steps you are taking. It's important that everyone on the team understand the rationale behind the metrics being used to judge success or failure of the project.

Communications management is about keeping everybody in the loop. Have you ever tried talking to someone in a really loud, crowded room? That's what running a project is like if you don't get a handle on communications. The communications planning process concerns defining the types of information you're going to deliver, to whom, the format for communicating the information and when. It turns out that 90% of a project manager's job is spent on communication so it's important to make sure everybody gets the right message at the right time.

The first step in defining your communication plan is figuring out what kind of communication your stakeholders need from the project so that they can make good decisions. This is called the communications requirements analysis. Your project will produce a lot of information; you don't want to overwhelm your stakeholders with all of it. Your job here is to figure out what they feel is valuable. Communicating valuable information doesn't mean you always paint a rosy picture. Communications to stakeholders may consist of either good news or bad news-the point is that you don't want to bury stakeholders in too much information but give them enough so that they're informed and can make appropriate decisions.

Communications technology has a major impact on how you can keep people in the loop. This examines the methods (or technology) used to communicate the information to, from and among the stakeholders. Methods of communicating can take many forms, such as written, spoken, e-mail, formal status reports, meetings, online databases, online schedules, project websites and so forth. You should consider several factors before deciding what methods you'll choose to transfer information. The timing of the information exchange or need for updates is the first factor. It's a lot easier for people to get information on their projects if it's accessible through a web site, than if all your information is passed around by paper memos. Do you need to procure new technology or systems, or are there systems already in place that will work? The technologies available to you will definitely figure into your plan of how you will keep everyone notified of project status and issues. Staff experience with the technology is another factor. Are there project team members and stakeholders experienced at using this technology, or will you need to train them? Finally, consider the duration of the project and the project environment. Will the technology you're choosing work throughout the life of the project or will it have to be upgraded or updated at some point? And how does the project team function? Are they located together or spread out across several campuses or locations?

The answers to these questions should be documented in the communication plan.

All projects require sound communication plan, but not all projects will have the same types of communication or the same methods for distrusting the information. The communication plan documents the types of information needs the stakeholders have, when the information should be distributed and how the information will be delivered.

The type of information you will typically communicate includes project status, project scope statements, and scope statement updates, project baseline information, risks, action items, performance measures, project acceptance and so on. What's important to know now is that the information needs of the stakeholders should be determined as early in the planning phase of the project management lifecycle as possible so that as you and your team develop project planning documents, you already know who should receive copies of them and how they should be delivered.

Believe it or not, we have officially completed the planning phase of the project management lifecycle. The project plan is the approved, formal, documented plan that's used to guide you throughout the project implementation phase. The plan is made up of all the processes of the planning phase. It is the map that tells you where you're going and how to perform the activities of the project plan during the project implementation phase. It serves several purposes; the most important of which is tracking and measuring project performance. The project plan is critical in all communications you'll have from here forward with the stakeholders, management, and customers. The project plan encompasses everything we talked about up to now and is represented in a formal document or collection of documents. This document contains the project scope, deliverables, assumptions, risks, WBS, milestones, project schedule, resources, communication plan, the project budget and any procurement needs. It becomes the baseline you'll use to measure and track progress against. It is also used to help you control the components that tend to stray away from the original plan so you can get them back on track.

The project plan is used as a communication and information tool for stakeholders, team members and the management team. They will use the project plan to review and gauge progress as well. Your last step in the planning phase is obtaining sign-off of the project plan from stakeholders, the sponsor and the management team. If they've been an integral part of the planning processes all along (and I know you know how important this is), obtaining sign-off of the project plan should simply be a formality.

PART
IV

Part IV - IMPLEMENTATION and CLOSING

Waning Moon Copyright   CC BY ChuckThePhotographer
www.flickr.com/photos/chuckthephotographer/2300242223/sizes/m/in/photostream/

After you have carefully planned your project, you will be ready to start the project implementation phase, the third phase of the project management life cycle. The implementation phase involves putting the project plan into action. It's here that the project manager will coordinate and direct project resources to meet the objectives of the project plan. As the project unfolds, it's the project manager's job to direct and manage each activity on the project, every step of the way. That's what happens in the implementation phase of the project lifecycle; you simply follow the plan you've put together and handle any problems that come up.

The implementation phase is where you and your project team actually do the project work to produce the deliverables. The word deliverable means anything your project delivers. The deliverables for your project include all of the products or services that you and your team are performing for the client, customer or sponsor including all the project management documents that you put together.

The steps undertaken to build each deliverable will vary depending on the type of project you are undertaking, and cannot therefore be described here in any real detail. For instance engineering and telecommunications projects will focus on using equipment, resources and materials to construct each project deliverable, whereas computer software projects may require the development and implementation of software code routines to produce each project deliverable. The activities required to build each deliverable will be clearly specified within the project requirements document and project plan accordingly.

Your job as project manager is to direct the work, but you need to do more than deliver the results. You also need to keep track of how well your team performed. The executing phase keeps the project plan on track with careful monitoring and control processes to ensure the final deliverable meets the acceptance criteria set by the customer. This phase is typically where approved changes are implemented.

Most often changes are identified through looking at performance and quality control data. Routine performance and quality control measurements should be evaluated on a regular basis throughout the implementation phase. Gathering reports on those measurements will help you determine where the problem is and recommend changes to fix it.

22.1 Change control

When you find a problem, you can't just make a change, because what if it's too expensive, or will it take too long? You will need to look at how it affects the triple constraint (time, cost, scope) and how they impact quality. You will then have to figure out if it is worth making the change. Change control is a set of procedures that let you make changes in an organized way.

Anytime you need to make a change to your plan, you need to start with a change request (Figure 22.1). This is a document that either you or the person making the request needs to create. Any change to your project needs to be documented so you can figure out what needs to be done, by when, and by whom.

Once the change request is documented, it is submitted to a change control board. A change control board is a group of people who consider changes for approval. Not every change control system has a board but most do. The change request could also be submitted to the project sponsor or management for review and approval. Putting the recommended changes through change control will help you evaluate the impact and update all the necessary documents. Not all changes are approved, but if the changes and repairs are approved, you send them back to the team to put them in place.

The implementation phase will use the most project time and resources and as a result, costs are usually the highest during the executing phase. Project managers will also experience the greatest confects over schedules in this phase. You may find as your monitoring your project, the actual time it is taking to do the scheduled work is taking longer than the amount of time you planned. If you evaluate the impact of the change and find that it won't have an impact on the project triple constraint, then you can make the change without going through change control.

When you absolutely have to meet the date and you are running behind, you can sometimes find ways to do activities more quickly by adding more resources to critical path tasks. That's called crashing. Crashing the schedule means adding resources or moving them around to shorten it. Crashing **always** costs more and doesn't always work. There's no way to crash a schedule without raising the overall cost of the project. So, if the budget is fixed and you don't have any extra money to spend, you can't use this technique.

Sometimes you've got two activities planned to occur in sequence, but you can actually do them at the same time. This is called fast-tracking the project. On a software project, you might do both your user acceptance testing (UAT) and your functional testing at the same time, for example. This is pretty risky. There's a good chance you might need to redo some of the work you have done concurrently. Crashing and fast tracking are schedule compression tools. Managing a schedule change means keeping all of your schedule documents up to date. That way, you will always be comparing your results to the correct plan.

After the deliverables have been physically constructed and accepted by the customer, a phase review is carried out to determine whether the project is complete and ready for closure.

Every project needs to end and that's what project completion is all about in the last phase of the project lifecycle. The whole point of the project is ~~that you need~~ to deliver what you promised. By making sure you delivered everything you said you would, you make sure that all stakeholders are satisfied and all acceptance criteria have been met. Once that happens, your project can finish.

Project completion is often the most often neglected phase of all the project lifecycles. Once the project is over, it's easy to pack things up, throw some files in a drawer, and start moving right into the initiation phase of the next project. Hold on. You're not done yet.

The key activity in project completion is gathering project records and disseminating information to formalize acceptance of the product, service or project as well as to perform project closure. As the project manager, you will need to review project documents to make certain they are up-to-date. For example, perhaps some scope change requests were implemented that changed some of the characteristics of the final product. The project information you are collecting during this phase should reflect the characteristics and specifications of the final product. Don't forget to update your resource assignments as well. Some team members will have come and gone over the course of the project. You need to double-check that all the resources and their roles and responsibilities are noted.

Once the project outcomes are documented, you'll request formal acceptance from the stakeholders or customer. They're interested in knowing if the product or service of the project meets the objectives the project set out to accomplish. If your documentation is up-to-date, you'll have the project results at hand to share with them.

23.1   Lessons learned

Project completion is also concerned with analyzing the project management processes to determine their effectiveness and to document lessons learned. Lessons learned are used to document the successes and failures of the project. As an example, lessons learned document the reasons why specific corrective actions were taken, their outcomes, the causes of performance variances, unplanned risks that occurred, mistakes that were made and could have been avoided and so on.

Unfortunately, sometimes projects do fail. There are things that can be learned from the failure of a project (as well from successful projects), and this information should be documented for future reference. Lessons learned can be some of the most valuable information you'll take away from any project. We can all learn from our experiences, and what better way to have more success on your next project than to review a similar past project's lessons learned document? But it is important not to forget the lessons learned.

23.2   Contract closure

Contracts come to a close just as projects come to a close. Contract closure is concerned with completing and settling the terms of the contract. It supports the project completion process because the contract closure process determines if the work described in the contract was completed accurately and satisfactorily. Keep in mind that not all projects are performed under contract so not all projects require the contract closure process. Obviously, this process applies only to those phases, deliverables or portions of the project that were performed under contract.

Contract closure updates the project records detailing the final results of the work on the project. Contracts may have specific terms or conditions for completion. You should be aware of these terms or conditions so that project completion isn't held up because you missed an important detail. If you are administering the contract yourself, be sure to ask your procurement department if there are any special conditions that you should be aware of so that your project team doesn't inadvertently delay contract project closure.

One of the purposes of contract closure process is to provide formal notice to the seller-usually in written form, that the deliverables are acceptable and satisfactory or have been rejected. If the product or service does not meet the expectations, the vendor will need to correct the problems before you issue a formal acceptance notice. Hopefully, quality audits have been performed during the course of the project and the vendor was given the opportunity to make corrections earlier in the process than the closing phase. It's not a good idea to wait until the very end of the project and then spring all the problems and issues on the vendor at once. It's much more efficient to discuss problems with your vendor as the project progresses because it provides the opportunity for correction when the problems occur.

If the product or service does meet the project's expectation and is acceptable, formal written notice to the seller is required indicating that the contract is complete. This is the formal acceptance and closure of the contract. It's your responsibility as the project manager to document the formal acceptance of the contract. Many times the provisions for formalizing acceptance and closing the contract are spelled out in the contract itself.

If you have a procurement department handling the contract administration, they will expect you to inform them when the contract is complete and will in turn follow the formal procedures to let the seller know the contract is complete. However, you'll still note the contract completion in your copy of the project records.

23.3   Releasing project team

Releasing project team members is not an official process. However, it should be noted that at the conclusion of the project, you will release your project team members, and they will go back to their functional managers or get assigned to a new project. You will want to keep their managers, or other project managers, informed as you get closer to project completion, so that they have time to adequately plan for the return of their employees. Let them know a few months ahead of time what the schedule looks like and how soon they can plan on using their employees on new projects. This gives the other managers the ability to start planning activities and scheduling activity dates.

23.4   Celebrate!

The project team should celebrate their accomplishments, and the project manager should officially recognize their efforts and thank them for their participation and officially close the project. A celebration helps team members formally recognize the project's end and brings closure to the work they've done. It also encourages them to remember what they've learned and to start thinking about how their experiences will benefit them and the organization during the next project (Figure 23.1).

Figure 23.1: Celebrate! Your project is over... at least until the next one.

Appendix A: Solutions to Exercises

Solution to Exercise 10.1

The floor and all furniture will be covered by drop cloths prior to painting each of the four office walls, not including the door, base-boards, crown molding, window frames, windows, light switches, or power outlets that will be masked with painters tape. The walls should be painted using Behr paint in Sapphire Berry #550C-2 in the semi-gloss sheen, using a nylon/polyester 3" inch wide brush and a vertical brush stroke. Two coats of paint should be applied, allowing 16 hours between coats, and the painting should be completed by Friday, November 7, including removal of all protective tape and drop cloths.

Of course, this assumes the contractor knows which office to paint!

Solution to Exercise 15.1

- Invitations
    - Create guest list
    - Print the invitations
    - Mail the invitations
    - Wait for RSVPs
- Food
    - Find caterer
    - Finalize the menu
    - Cater the wedding
- Bridal
    - Shop for dress
    - Shop for shoes
    - Choose the bouquet
    - Tailoring and fitting

In pictorial format:

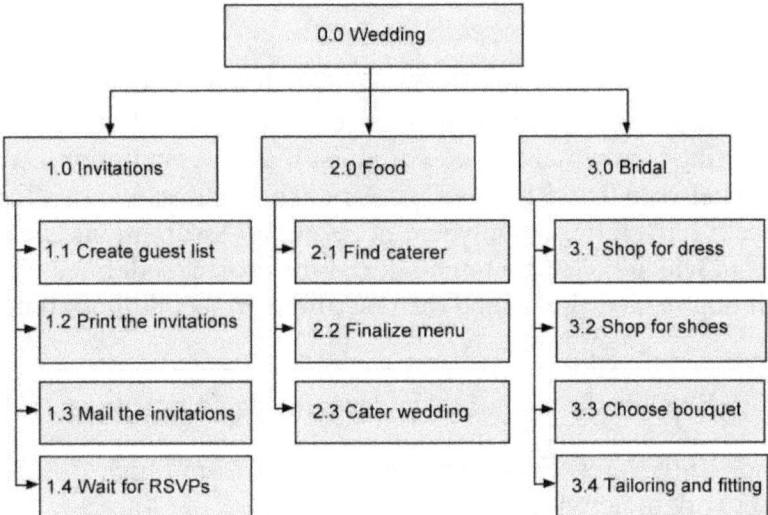

Illustration from Barron & Barron Project Management for Scientists and Engineers, http://cnx.org/content/col11120/1.4/

Solutions to Exercises in Chapter 16

• Solution to Exercise 16.1	• Alternative analysis.
• Solution to Exercise 16.2	• Published estimating data.
• Solution to Exercise 16.3	• Expert judgment.
• Solution to Exercise 16.4	• Project management software
• Solution to Exercise 16.5	• Bottom-up estimating
• Solution to Exercise 16.6	• Alternative analysis
• Solution to Exercise 16.7	• Expert judgment.
• Solution to Exercise 16.8	• Expert judgment.
• Solution to Exercise 16.9	• Parametric estimating.
• Solution to Exercise 16.10	• Analogous estimating.

Assumption - There may be external circumstances or events that must occur for the project to be successful (or that should happen to increase your chances of success). If you believe that the probability of the event occurring is acceptable, you could list it as an assumption. An assumption has a probability between 0 and 100%. That is, it is not impossible that the event will occur (0%) and it is not a fact (100%). It is somewhere in between. Assumptions are important because they set the context in which the entire remainder of the project is defined. If an assumption doesn't come through, the estimate and the rest of the project definition may no longer be valid.

BAC - Budget at completion (BAC) is the sum of all budgets allocated to a project.

Backward pass - Calculation of the latest finish times by working from finish-to-start for the uncompleted portion of a network of activities.

Balanced matrix - An organizational matrix where functions and projects have the same priority.

Bar chart - A view of the project schedule that uses horizontal bars on a time scale to depict activity information; frequently called a Gantt chart.

Baseline - The value or condition against which all future measurements will be compared.

Baseline cost - The amount of money an activity was intended to cost when the baseline plan was established.

Baseline dates - Original planned start and finish dates for an activity that is used to compare with current planned dates to determine any delays; also used to calculate budgeted cost of work scheduled in earned value analysis

Baseline plan - The original plan (for a project, a work package, or an activity), plus or minus approved changes. Usually used with a modifier, e.g., cost baseline, schedule baseline, performance measurement baseline, etc.

Best practices - Techniques that agencies may use to help detect problems in the acquisition, management, and administration of service contracts; best practices are practical techniques gained from experience that have been shown to produce best results.

Beta testing - Pre-release testing in which a sampling of the intended customer base tries out the product.

Bottom-up cost estimate - The approach to making a cost estimate or plan in which detailed estimates are made for every task shown in the work breakdown structure and then summed to provide a total cost estimate or plan for the project.

Brainstorming - The unstructured and dynamic generation of ideas by a group of people where anything and everything is acceptable, it is particularly useful in generating a list of potential project risks.

Budget - Generally refers to a list of all planned expenses and revenues.

Budgeted cost of work performed (BCWP) - Measures the budgeted cost of work that has actually been performed rather than the cost of work scheduled

Budgeted cost of work scheduled (BCWS) - The approved budget that has been allocated to complete a scheduled task, or work breakdown structure (WBS) component, during a specific time period

Business analysis - The set of tasks, knowledge, and techniques required to identify business needs and determine solutions to business problems. Solutions often include a systems development component, but may also consist of process improvement or organizational change.

Business area - The part of the organization containing the business operations affected by a program or project

Business case - A document developed towards the end of the concept phase, to establish the merits and desirability of the project and justification for further project definition.

Business needs - The requirements of an enterprise to meet its goals and objectives.

Business operations - The ongoing recurring activities involved in the running of a business for the purpose of producing value for the stakeholders. They are contrasted with project management, and consist of business processes.

Business process - A collection of related, structured activities or tasks that produce a specific service or product (serve a particular goal) for a particular customer or customers. There are three types of business processes: management processes, operational processes, and supporting processes.

Case study - A research method that involves an in-depth, longitudinal examination of a single instance or event: a case. It provides a systematic way of looking at events, collecting data, analyzing information, and reporting the results.

Champion - An end-user representative, often seconded into a project team. Someone who acts as an advocate for a proposal or project.

Change control - A general term describing the procedures used to ensure that changes (normally, but not necessarily, to IT systems) are introduced in a controlled and coordinated manner. Change control is a major aspect of the broader discipline of change management.

Change management - The formal process through which changes to the project plan are approved and introduced.

Change order - A document that authorizes a change in some aspect of the project.

Change request - A request needed to obtain formal approval for changes to the scope, design, methods, costs, or planned aspects of a project. Change requests may arise through changes in the business or issues in the project. Change requests should be logged, assessed and agreed-on before a change to the project can be made.

Child activity - Subordinate task belonging to a parent task existing at a higher level in the work breakdown structure.

Client/customers - The person or group that is the direct beneficiary of a project or service is the client/customer. These are the people for whom the project is being undertaken (indirect beneficiaries are stakeholders). In many organizations, internal beneficiaries are called clients and external beneficiaries are called customers, but this is not a hard and fast rule.

Completion - The completion of all work on a project.

Communication plan - A statement of the project's stakeholders' communication and information needs.

Concept phase - The first phase of a project in the generic project lifecycle, in which the need is examined, alternatives are assessed, the goals and objectives of the project are established, and a sponsor is identified.

Confidence level - A level of confidence, stated as a percentage, for a budget or schedule estimate. The higher the confidence level, the lower the risk.

Conflict management - Handling of conflicts between project participants or groups in order to create optimal project results.

Conflict resolution - The solution to a problem, using one of five methods in particular: confrontation, compromise, smoothing, forcing and withdrawal.

Constraints - Limitations that are outside the control of the project team and need to be managed around. They are not necessarily problems. However, the project manager should be aware of constraints because they represent limitations within which the project must be executed. Date constraints, for instance, imply that certain events (perhaps the end of the project) must occur by certain dates. Resources are almost always a constraint, since they are not available in an unlimited supply.

Contingencies - The planned allotment of time and cost for unforeseeable elements in a project. Including contingencies will increase the confidence of the overall project.

Control - The process of comparing actual performance with planned performance, analyzing the differences, and taking the appropriate corrective action.

Costs - The cost value of project activity.

Costs budgeting - The allocation of cost estimates to individual project components.

Cost overrun - The amount by which actual costs exceed the baseline or approved costs.

Crashing - The process of reducing the time it takes to complete an activity by adding resources.

Critical - An activity or event that, if delayed, will delay some other important event, commonly the completion of a project or a major milestone in a project.

Critical path - The critical path is the sequence of activities that must be completed on-schedule for the entire project to be completed on schedule. It is the longest duration path through the work plan. For example, if an activity on the critical path is delayed by one day, the entire project will be delayed by one day (unless another activity on the critical path can be accelerated by one day).

Critical path method (CPM) - A mathematically based modeling technique for scheduling a set of project activities.

Critical chain project management (CCPM) - A method of planning and managing projects that puts more emphasis on the resources required to execute project tasks.

Critical success factors - The key factors that are deemed critical to the success of the project. The nature of these factors will govern the response to conflicts, risks and the setting of priorities.

Culture - A person's attitudes arising out of their professional, religious, class, educational, gender, age and other backgrounds.

Customer - See client.

Deliverable - *A deliverable is any tangible outcome that is produced by the project. All projects create deliverables. These can be documents, plans, computer systems, buildings, aircraft, etc. Internal deliverables are produced as a consequence of executing the project and are usually needed only by the project team. External deliverables are those that are created for clients and stakeholders. Your project may create one or many deliverables.*

Dependency - Dependencies are the relationships between activities whereby one activity must do something (finish-to-start) before another activity can do something (start-to-finish).

Duration - The duration of a project's terminal element is the number of calendar periods it takes from the time the implementation of an element starts to the moment it is completed.

Earned value management (EVM) - A project management technique for measuring project progress in an objective manner, measuring scope, schedule, and cost combined in a single integrated system.

Earned schedule (ES) - An extension to earned value management (EVM), which renames two traditional measures to indicate clearly they are in units of currency or quantity, not time.

Estimation - The processes of making accurate estimates using the appropriate techniques

Event chain diagram - A diagram that shows the relationships between events and tasks and how the events affect each other

Event chain methodology - An uncertainty modeling and schedule network analysis technique that is focused on identifying and managing events and event chains that affect project schedules

Float - The amount of time that a task in a project network can be delayed without causing a delay to subsequent tasks and/or the project completion date

Functional manager - The person you report to within your functional organization. Typically, this is the person who does your performance review. The project manager may also be a functional manager, but he or she does not have to be. If your project manager is different from your functional manager, your organization is probably utilizing matrix management.

Gantt, Henry - An American mechanical engineer and management consultant who developed the Gantt chart in the 1910's

Gantt chart - A Gantt chart is a bar chart that depicts activities as blocks over time. The beginning and end of the block correspond to the beginning and end-date of the activity.

Goal - An objective that consists of a projected state of affairs which a person or a system plans or intends to achieve or bring about. A personal or organizational desired end-point in some sort of assumed development. Many people endeavor to reach goals within a finite time by setting goals.

Goal setting - Involves establishing specific, measurable and time-targeted objectives.

Graphical evaluation and review technique (GERT) - A network analysis technique that allows probabilistic treatment of both network logic and activity duration estimation

Hammock activity - A schedule or project planning term for a grouping of subtasks that hang between two end dates to which they are tied or the two end events to which they are fixed

ISO 10006 - A guideline for quality management in projects, it is an international standard developed by the International Organization for Standardization.

Issue - An issue is a major problem that will impede the progress of the project that can't be resolved by the project manager and project team without outside help. Project managers should proactively deal with issues through a defined issues management process.

Kickoff meeting - The first meeting with the project team and the client of the project

Level of effort (LOE) - A support-type activity which doesn't lend itself to measurement of a discrete accomplishment. Examples may be project budget accounting, customer liaison, etc.

Life cycle - The process used to build the deliverables produced by the project. Every project has an inception, a period during which activities move the project toward completion, and a termination (either successful or unsuccessful).

Management - Management comprises planning, organizing, staffing, leading or directing, and controlling an organization (a group of one or more people or entities) or effort for the purpose of accomplishing a goal.

Management process - A process of planning and controlling the performance or implementation of any type of activity

Motivation - The reasons that entice one to engage in a particular behavior

Milestone - A scheduling event that signifies the completion of a major deliverable or a set of related deliverables. A milestone, by definition, has a duration of zero and no effort. There is no work associated with a milestone. It is a flag in the work plan to signify that some other work has been completed. Usually, a milestone is used as a project checkpoint to validate how the project is progressing. In many cases there is a decision, such as validating that the project is ready to proceed further, that needs to be made at a milestone.

Objective - A concrete statement that describes what the project is trying to achieve. The objective should be written at a low level so that it can be evaluated at the conclusion of a project to see whether it was achieved. Project success is determined based on whether or not the project objectives were achieved. A technique for writing an objective is to make sure it is specific, measurable, acceptable, realistic, and time-based (SMART).

Operations management - An area of business that is concerned with the production of good quality goods and services, and involves responsibility for ensuring that business operations are efficient and effective. It is the management of resources, the distribution of goods and services to customers, and the analysis of queue systems.

Organization - A social arrangement that pursues collective goals, controls its own performance, and has a boundary separating it from its environment.

Planning - Processes that involve formulating and revising project goals and objectives and creating the project management plan that will be used to achieve the goals the project was undertaken to address. Planning involves determining alternative courses of action and selecting from among the best of those to produce the project's goals.

Process - An ongoing collection of activities, with inputs, outputs and the energy required to transform inputs to outputs.

Program - The umbrella structure established to manage a series of related projects. The program does not produce any project deliverables; the project teams produce them all. The purpose of the program is to provide overall direction and guidance, to make sure the related projects are communicating effectively, to provide a central point of contact and focus for the client and project teams, and to determine how individual projects should be defined to ensure that all the work gets completed successfully.

Program management - The process of managing multiple, ongoing, inter-dependent projects. An example would be that of designing, manufacturing and providing support infrastructure for an automobile manufacturer.

Program manager - The person with the authority to manage a program. (Note that this is a role. The program manager may also be responsible for one or more projects within the program.) The program manager leads the overall planning and management of the program. All project managers within the program report to the program manager.

Project - A project is a temporary endeavor undertaken to accomplish a unique product or service with a defined start and end point and specific objectives that, when attained, signify completion.

Project definition (project charter) - Before you start a project, it is important to know the overall objectives of the project, as well as the scope, deliverables, risks, assumptions, project organization chart, etc. The project definition (or project charter) is the document that holds this relevant information. The project manager is responsible for creating the project definition. The document should be approved by the sponsor to signify that the project manager and the sponsor are in agreement on these important aspects of the project.

Project manager - The person with the authority to manage a project. The project manager is 100 percent responsible for the processes used to manage the project. He or she also has people management responsibilities for team members, although this is shared with the team member's functional manager. The processes a project manager might perform include defining the work, building the work plan and budget, managing the work plan and budget, scope management, issues management, risk management, etc.

Project management - The application of knowledge, skills, tools, and techniques applied to project activities in order to meet or exceed stakeholder needs and expectations from a project.

Project management body of knowledge (PMBOK) - The sum of knowledge within the profession of project management that is standardized by ISO.

Project management professional - A certificated professional in project management (PMP).

Project management software - Software that includes scheduling, cost control and budget management, resource allocation, collaboration tools, communication, quality management, and documentation systems, which are used to deal with the complexity of large projects.

Project phase - A major logical grouping of work on a project, it also represents the completion of a major deliverable or set of related deliverables. On an IT development project, logical project phases might be planning, analysis, design, construction (including testing), and implementation.

Project plan - A formal, approved document used to guide both project implementation and project control. The primary uses of the project plan are to document planning assumptions and decisions, facilitate communication among stakeholders, and document approved scope, cost, and schedule baselines. There are two types of project plans: summary or detailed.

Project planning - The use of schedules such as Gantt charts to plan and subsequently report progress within the project environment.

Project team - Full and part-time resources assigned to work on project deliverables. They are responsible for understanding the work to be completed; completing assigned work within the budget, timeline, and quality expectations; informing the project manager of issues, scope changes, and risk and quality concerns; and proactively communicating status and managing expectations.

Quality - The standards and criteria to which the project's products must be delivered for them to perform effectively. First, the product must perform to provide the functionality expected, solve the problem, and deliver the benefit and value expected of it. It must also meet other performance requirements, or service levels, such as availability, reliability and maintainability, and have acceptable finish and polish. Quality on a project is controlled through quality assurance (QA), which is the process of evaluating overall project performance on a regular basis to provide confidence that the project will satisfy the relevant quality standards.

Requirements - Descriptions of how a product or service should act, appear, or perform and which generally refer to the features and functions of the project deliverables being built. Requirements are considered to be a part of project scope. High-level scope is defined in your project definition (charter). The requirements form the detailed scope. After your requirements are approved, they can be changed through the scope change management process.

Resources - Resources are the people, equipment, and materials needed to complete the work of the project.

Risk - There may be potential external events that will have a negative impact on your project if they occur. Risk refers to the combination of probability that these events will occur and the impact on the project if they do. If the combination of the probability of the event and the impact to the project is too high, you should identify the potential event as a risk and put a proactive plan in place to manage it.

Risk management planning - Determines how risks will be managed for a project and describes how risks are defined, monitored, and controlled throughout the project.

Schedule development - Calculates and prepares the schedule of project activities which becomes the schedule baseline. It determines activity start and finish dates, finalizes activity sequences and durations, and assigns resources to activities.

Scope - Describes the boundaries of the project. It defines what the project will deliver and what it will not.

Scope creep - Refers to uncontrolled changes in a project's scope. This phenomenon can occur when the scope of a project is not properly defined, documented, or controlled. It is generally considered a negative occurrence to be avoided.

Six sigma - A business management strategy, originally developed by Motorola, that today enjoys widespread application in many sectors of industry.

Sponsor (executive sponsor and project sponsor) - The person who has ultimate authority over the project. The executive sponsor provides project funding, resolves issues and scope changes, approves major deliverables, and provides high-level direction. He or she also champions the project within the organization. Depending on the project and the organizational level of the executive sponsor, he or she may delegate day-to-day tactical management to a project sponsor. If assigned, the project sponsor represents the executive sponsor on a day-to-day basis and makes most of the decisions requiring sponsor approval. If the decision is large enough, the project sponsor will take it to the executive sponsor.

Stakeholder - Specific people or groups who have a stake in the outcome of the project are stakeholders. Normally stakeholders are from within the company and may include internal clients, management, employees, administrators, etc. A project can also have external stakeholders, including suppliers, investors, community groups, and government organizations.

Steering committee - This is usually a group of high-level stakeholders who are responsible for providing guidance on overall strategic direction. They don't take the place of a sponsor, but help spread the strategic input and buy-in to a larger portion of the organization. The steering committee is especially valuable if your project has an impact on multiple organizations because it allows input from those organizations into decisions that affect them.

Systems development lifecycle (SDLC) - Any logical process used by a systems analyst to develop an information system, including requirements, validation, training, and user ownership. An SDLC should result in a high-quality system that meets or exceeds customer expectations, within time and cost estimates, works effectively and efficiently in the current and planned information technology (IT) infrastructure, and is cheap to maintain and cost-effective to enhance.

Task - An activity that needs to be accomplished within a defined period of time

Task analysis - The analysis or breakdown of exactly how a task is accomplished, including what sub-tasks are required.

Timeline - A graphical representation of a chronological sequence of events, also referred to as a chronology. It can also mean a schedule of activities, such as a timetable.

Triple constraint - Called the scope triangle or quality triangle, triple constraint illustrates the relationship between three primary forces in a project: scope, time and cost. Project quality is sometimes considered the fourth constraint or included in scope.

Work - The amount of effort applied to produce a deliverable or accomplish a task (a terminal element).

Work breakdown structure (WBS) - A task-oriented, family tree of activities which organizes, defines and graphically displays the total work to be accomplished in order to achieve the final objectives of a project. A system for sub-dividing a project into manageable work packages.

Work package - A deliverable at the lowest level of a work breakdown structure (WBS), they are a group of related tasks defined at the same level within the WBS.

Work plan (schedule) - Allows the project manager to identify the work required to complete the project and to monitor the work to determine whether the project is on schedule. It describes the activities required, the sequence of the work, who is assigned to the work, an estimate of how much effort is required, when the work is due, and other information of interest to the project manager.

Appendix C: Attributions and Bibliography

Based on Project Management for Scientists and Engineers by
Merrie Barron and Andrew Barron
http://cnx.org/content/col11120/1.4/

Significant contributions from the following sources

Maura Irene Jones, Career Descriptions in Chapter 1
Several photographs Copyright © 2011 by Maura Irene Jones
Creative Commons Attribution 3.0 CC BY
Attribution URL http://www.linkedin.com/in/maurajones

Randy Fisher, Chapter 12
(a subset of Organization Management and Development at http://wikieducator.org/OMD/Culture_PM)

Rekha Raman, Microsoft Word template and formatting

Ali Daimee, syllabus and more

Mike Milos, syllabus

Shuly Cooper, reviews

Bob Sawyer, Project Manager for the Saylor Foundation proposal

Victor Cesena, Project Manager for the bound textbook

Jim Huether, Program Manager

Jacky Hood, Managing Editor

Dalvinder Singh, Copy Editor

Daria Hemming, Copy Editor

References in the Barron & Barron textbook:

CHAOS 2009 Summary and EPPM Study. The Standish Group, Boston, MA (2009).

J. Westland. *The Project Management Lifecycle*. Kogan Page Limited (2006).

J. Green and A. Stellman. *Head First PMP*. O'Reilly Media, CA (2007).

K. Heldman, C. Baca, and P. Jansen. *Project Manager Professional Study Guide.*, Wiley Publishing, Inc. NJ (1995).

Reference for Chapter 1: *Occupational Outlook Handbook* (OOH), 2010-11 Edition

References for Chapter 12

Casmir, Michael J. (2008). *Culture and the Changing Environment.*

Connor, P. E., & Lake, L. K. (1988). *Managing organizational change.* New York: Praeger.

Deal, T. E. (1985). *Cultural change: Opportunity, silent killer, or metamorphosis?* In R. Retrieved October 28, 2011 from http://cnx.org/content/m13465/latest/

Dodd, C.H. (1995). *Dynamics of Intercultural Communication*, 4th ed., Duguque, 1A, Brown and Benchmark.

Lindahl, Robert. *The Role of Organizational Climate and Culture in the School Improvement Process: A Review of the Knowledge Base*, in Connexions. Retrieved October 28, 2011 from http://cnx.org/content/m13465/latest/

Fontaine, G. (2005). *A Self-Organization Perspective on the Impact of Local verses Global Assignment Strategies and Knowledge Building. International Journal of Diversity in Organizations*, Communities and Nations, 5(1), 57-66.

Fontaine, Gary (2005). *Motivations for Going International: Profiles of Asian and American Foreign Study Students, Cross-Cultural Management Students and Global Managers*, International Journal of Management.

Hall, Edward T. (originally published 1959; 1973). *The Silent Language.* Anchor.

Hall, G. E., & Hord, S. M. (2001). Implementing change: Patterns, principles, and potholes. Needham Heights, MA: Allyn & Bacon.

Johnson, D. & Johnson, F. (2005). *Joining Together: Group Theory and Group Skills*, Ninth Edition. Boston: Allyn and Bacon.

Merriam-Webster Online Dictionary (2011). Culture. Retrieved on October 28, 2011 from http://www.merriam-webster.com/dictionary/culture

Owens, R. G. (2004). *Organizational behavior in education: Adaptive leadership and school reform* (8th ed.). Boston: Allyn & Bacon.

Pelton, Robert Young (2003). *The World's Most Dangerous Places*, Collins; 5th Rev edition.

Rousseau, D. M. (1990). *Assessing organizational culture: The case for multiple methods.* In B. Schneider (Ed.), Organizational climate and culture (pp. 153 - 192). San Francisco: Jossey-Bass.

Schein, E. H. (1984, Summer). *Suppose we took culture seriously.* Academy of Management OD Newsletter, 2ff.

Schein, E. H. (1985a). *How culture forms, develops, and changes.* In R. H. Kilman, M.J.

Saxton, & R. Serpa (Eds)., *Gaining control of the corporate culture* (pp. 17-43). San Francisco: Jossey-Bass.

Schein, E. H. (1985b). *Organizational culture and leadership*. San Francisco: Jossey-Bass.

Schein, E. H. (1992). *Organizational culture and leadership* (2nd ed.). San Francisco: Jossey-Bass.

Schein, E.H. (1999). *The corporate culture survival guide: Sense and nonsense about culture change*. San Francisco: Jossey-Bass.

Shebib, Bob (2003). Bob. *Choices: Interviewing and Counselling Skills for Canadians*, 2nd edition, Pearson Education Canada Inc.

Thompson, K. R., & Luthans, F. (1990). *Organizational culture: A behavioral perspective*. In B. Schneider (Ed.), Organizational climate and culture (pp. 319-344). San Francisco: Jossey-Bass.

Trompenaars, F. & Hampden-Turner, C. (1998). *Riding the Waves of Culture: Understanding Cultural Diversity in Global Business*. Irwin.

The Cain Project in Engineering and Professional Communication. *Guide to Communication and Corporate Culture.* Connexions. Retrieved October 28, 2011 from http://cnx.org/content/m17116/latest/

Wheatley, M. (2006). *Leadership and the New Science: Discovering order in a chaotic world*. San Francisco, Berrett-Koehler

Wilkins, A. L., & Patterson, K. J. (1985). *You can't get there from here: What will make culture-change projects fail*. In R. H. Kilman, M. J. Saxton, & R. Serpa (Eds)., *Gaining control of the corporate culture* (pp. 262-291). San Francisco: Jossey-Bass.

www.ingramcontent.com/pod-product-compliance
Lightning Source LLC
Chambersburg PA
CBHW080659190526
45169CB00006B/2185